1936 INSTRUCTION

ON

THE TACTICAL EMPLOYMENT OF LARGE UNITS

The Tower and Anchor Military Archive of World War II

Published or forthcoming:

Translations of French World War II Military Manuals:

French World War II Armored Doctrine

1926 Artillery Maneuver Regulations, Part 2, The Artillery in Combat

1936 Instruction on the Tactical Employment of Large Units

1938 Infantry Regulations

Translations of Italian World War II Military Manuals:

1936 Norms for Division Combat

1937 Regulations on Artillery Training, Vol. III Tactical Deployment and Training, Part 1 Artillery in Combat

1938 Norms for the Employment of Tank Units

1939 Regulations on Infantry Training, Vol. II Tactical Employment and Training

Related publications:

A Glossary of World War II Military Terms: Compiled from Official U.S. War Department and British War Office Documents

FRANCE
MINISTÈRE DE LA DÉFENSE NATIONALE ET DE LA GUERRE

1936 INSTRUCTION

ON

THE TACTICAL EMPLOYMENT OF LARGE UNITS

Translated from
Instruction sur l'emploi tactique des Grandes Unités (1936)

Translation and Introduction by
Nicholas G. Forte

TOWER & ANCHOR BOOKS
PENSACOLA, FLORIDA

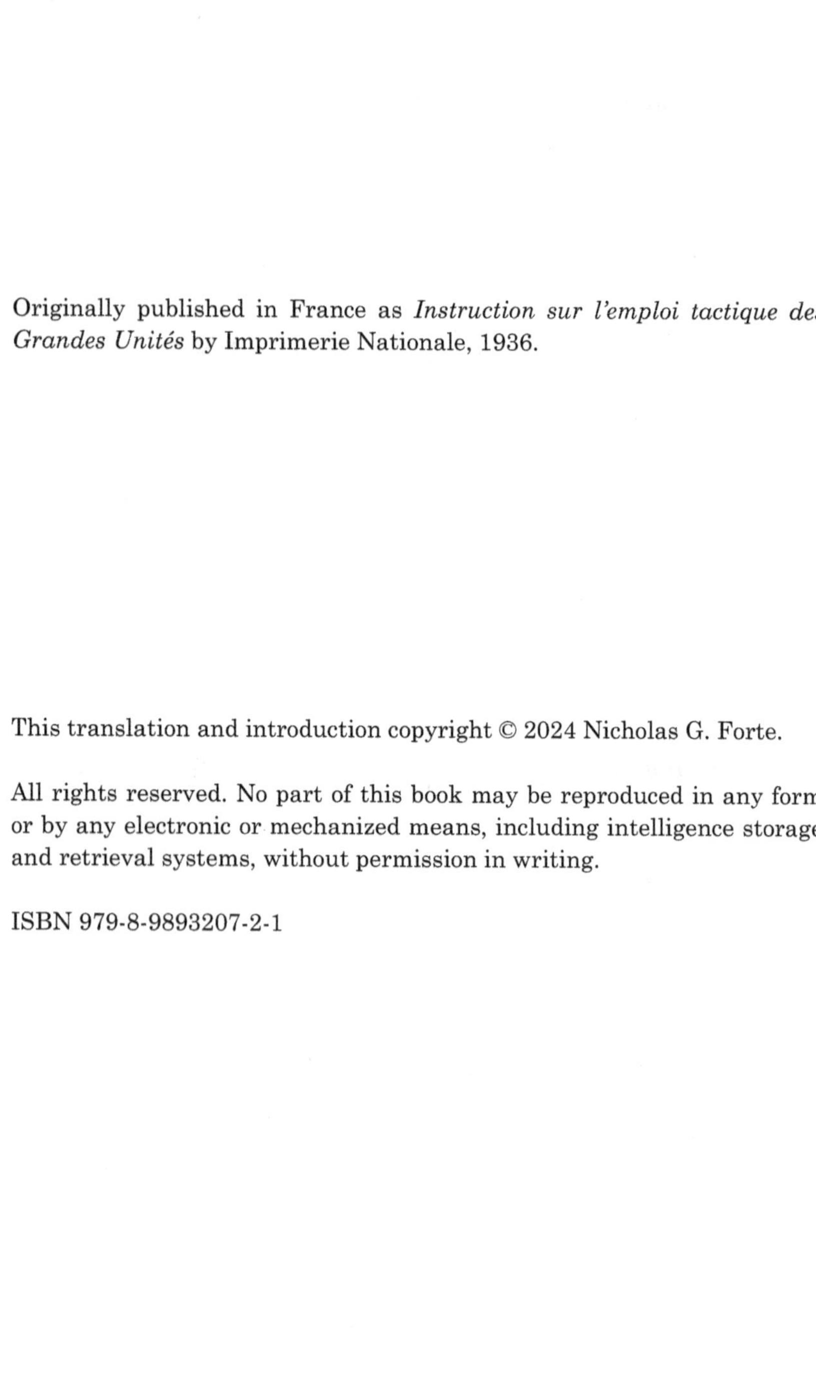

Originally published in France as *Instruction sur l'emploi tactique des Grandes Unités* by Imprimerie Nationale, 1936.

ISBN 979-8-9893207-2-1

INTRODUCTION

Published in 1936, the *Instruction on the Tactical Employment of Large Units* was referred to by French army officers as the "Bible" of their military doctrine. Intended for the commanders of divisional units and higher, it was the central document on which the French Army's tactical regulations and instructions were based at the start of the Second World War.

French military doctrine fundamentally rested on three pillars: the importance of firepower, the primacy of defense, and the methodical battle (centralized control). These principles, which emerged from the experiences of the First World War, defined the modern battlefield for the French. The French, however, did not entirely abandon the idea of the offensive. The Instruction stated that only offensive action could yield decisive results, but the French believed that overwhelming superiority in artillery was essential for any successful offensive.

The *Instruction* attempted to address the challenges posed by technological advances, but the French High Command failed to grasp how fully these advances would revolutionize the battlefield. Despite the common criticism of the French for dispersing their tanks in "penny packets," their doctrine did contemplate the mass use of tanks but mistakenly thought that antitank guns would play the same role against tanks in a future war as machine guns did against infantry in the First World War. The French also failed to recognize the potential of aviation to substitute for massed artillery during an attack. This resulted in an underestimation of the immediacy of the threat posed by the German panzer divisions that French intelligence had detected pouring through the Ardennes.

The most glaring error in the French doctrine, however, was its continued emphasis on the "methodical battle," which centralized control to higher commands and deprived lower commands of the ability to take local initiatives. German doctrine, in contrast, relied on higher commands to set goals or objectives while giving subordinates the initiative to determine the best means to achieve them. This difference allowed the Germans to act more swiftly and effectively, getting inside the French decision loop.

*
* *

The contribution of France's doctrine to its military disaster in 1940 has been widely debated. While French doctrine certainly had its flaws, an argument can be made that it was never thoroughly tested. Don Alexander (1996) and Robert Doughty (1990) argue that an erroneous strategy, rather than outdated French tactics, was the primary cause of the German victory on the Meuse. They criticized General Gamelin, the chief-of-staff and commander-in-chief of the French Army, for his gamble in committing the Seventh Army — his strategic reserve — in a futile attempt to link up with the Dutch army at Breda. Gamelin's decision to commit to the Breda maneuver was made in the face of strong objections from other senior French generals, including Georges, Billotte, Giraud, and Vuillemin.

Furthermore, the French kept an excessive number of divisions behind the Maginot Line, whose permanent fortifications, according to the *Instruction*, should have allowed a "rigorous economy of forces." These two strategic errors deprived the two French armies defending the Meuse of both the number of divisions required by doctrine to hold their section of the line and the reserves to counter any breach that did occur by progressively sealing the breach in the line (known as the *colmatage*) followed by a counterattack. Doughty argues that "If the seven divisions of the Seventh Army had been available on the 13th and 14th [of May] and had been committed in front of the German forces, the entire course of the war may have been different."

However, other historians argue that French military doctrine was too flawed to counter the more mobile battlefield opened up by the advances in motorization, mechanization, and aviation and was the major factor that led to the defeat. For example, Elizabeth Kier (1997) contends that the French doctrine, with its rigid emphasis on a defensive posture and the methodical battle, was a product of domestic political considerations rather than strictly military ones, which resulted in a doctrine that was ill-suited to counter the dynamic and flexible German blitzkrieg tactics. Similarly, Richard Shuster (2023) argues that French doctrine heavily influenced the Allied strategy, making it overly focused on defensive preparations and leaving it without a coherent plan for offensive operations, making the Allies vulnerable to rapid German advances.

Whether the flaws in French doctrine were so serious as to make their 1940 defeat inevitable, even without the strategic errors made by their commanders, is still an open question. Whichever view one holds, I hope that this English translation of the *Instruction* will provide an important resource for historians and students of military history to more fully understand the military operations during the opening days of World War II.

*
* *

For the translation of technical military terms, I consulted the U.S. War Department's *TM 20-205 Dictionary of United States Army Terms* (1944), *TM 30-253 Military Dictionary: English-French, French-English* (1943), *TM 30-502 French Military Dictionary: English-French, French-English* (1944), and Cornélis de Witt Wilcox's *A French-English Military Technical Dictionary* (1917).

Nicholas G. Forte

FURTHER READING

Alexander, Don W. "Repercussions of the Breda Variant." *French Historical Studies* 8, No. 3 (1974): 459–88.

Doughty, Robert A. *The Breaking Point: Sedan and the Fall of France, 1940.* Hampden, Conn.: Archon Books, 1990.

——. "The French Armed Forces, 1918-1940," *Military Effectiveness, Vol. 2: The Interwar Period.* Ed. by Allan R. Millet and Williamson Murray. Boston: Allen & Unwin, 1988.

——. *The Seeds of Disaster: The Development of French Army Doctrine, 1919-1939.* Hamden, Conn., Archon Books, 1985.

Gamelin, Gen. Maurice. *Server.* 3 vols. Paris: Plon. 1946-1947.

House, Jonathan M. *Combined Arms Warfare in the Twentieth Century.* Lawrence: University Press of Kansas, 2001.

Jordan, Nicole. "Strategy and Scapegoatism: Reflections on the French National Catastrophe, 1940." *Historical Reflections / Réflexions Historiques* 22, No. 1. (1996); 11-32.

Kier, Elizabeth. *Imagining War: French and British Military Doctrine between the Wars.* Princeton, N.J.: Princeton University Press, 1997.

Kiesling, Eugenia C. *Arming Against Hitler: France and the Limits of Military Planning* (Modern War Studies). Lawrence: University Press of Kansas, 1996.

Shuster, Richard J. "Trying Not to Lose It": The Allied Disaster in France and the Low Countries, 1940," *Journal of Advanced Military Studies* 14, No. 1 (Spring 2023); 272-290.

MINISTÈRE DE LA DÉFENSE NATIONALE ET DE LA GUERRE

INSTRUCTION

SUR

L'EMPLOI TACTIQUE DES GRANDES UNITÉS

PARIS

IMPRIMERIE NATIONALE

1936

(Facsimile of the 1936 title page)

This instruction repeals and replaces the PROVISIONAL INSTRUCTION ON THE TACTICAL EMPLOYMENT OF LARGE UNITS of October 6, 1921.

TABLE OF CONTENTS.

CHAPTER ONE.

CHAPTER II.

CHAPTER III.

CHAPTER IV.

CHAPTER V.

TITLE V.

THE BATTLE.

CHAPTER ONE.

CHAPTER II.

CHAPTER III.

CHAPTER IV.

TITLE VI.

THE ARMY IN BATTLE.

CHAPTER ONE.

CHAPTER II.

TITLE VII.

THE ARMY CORPS IN BATTLE.

CHAPTER ONE.

CHAPTER II.

TITLE VIII.

THE INFANTRY DIVISION IN COMBAT.

CHAPTER ONE.

CHAPTER II.

CHAPTER III.

TITLE IX.

GENERAL POINTS ON THE USE OF LARGE CAVALRY UNITS.

CHAPTER ONE.

CHAPTER II.

CHAPTER III.

CHAPTER IV.

TITLE X.

GENERAL INFORMATION ON THE USE OF MOTORIZED LARGE UNITS.

SINGLE CHAPTER.

TITLE XI.

SPECIAL CASES OF USE OF LARGE UNITS.

SINGLE CHAPTER.

TITLE XII.

OPERATION OF THE SERVICES.

CHAPTER ONE.

CHAPTER II.

CHAPTER III.

CHAPTER IV.

CHAPTER V.

TITLE XIII.

PREFACE.

Respecting the international commitments to which *France* has subscribed, the French Government will endeavor, at the beginning of a war and in agreement with the Allies, to obtain from enemy Governments the commitment not to use combat gases as a weapon of war. If this undertaking is not obtained, it will reserve the right to act according to the circumstances.

REPORT TO THE MINISTER.

The Drafting Commission of the Provisional Instruction of 1921 had proposed, in its report to the Minister, that the conditions for the tactical employment of the Large Units should be determined based on the lessons of the war, which were still ingrained in everyone's mind.

While acknowledging the significant progress made since then in combat and transport capabilities, the Drafting Committee of the present Instruction nevertheless considered that this technical progress did not significantly alter the essential rules established by its predecessors in the tactical field.

It, therefore, recognized that the body of doctrine, objectively established immediately after victory by eminent leaders who had just exercised high commands, should remain the charter of the tactical employment of our Large Units.

Therefore, given the evolution of the matériel and the resulting organization of the army, the new Instruction has proposed:

— to define the possibilities of modern means;

— to regulate the conditions for their implementation in battle;

— lay down general rules for the conduct of newly created Large Units (motorized and mechanized);

— to supplement the 1921 Instruction on several points.

NEW IDEAS AND THEIR CONSEQUENCES.

In less than fifteen years, new and important matériel means have been created or developed in the armies, soon followed, moreover, by achievements intended to neutralize their effects.

The implementation of these new factors was aimed in particular at:

— organization of fortified fronts;

— the creation of motorized units as well as mechanized units and the appearance of antitank weapons that responded to them;

— the increase in the power of the air forces, accompanied by a significant development of antiaircraft defense;

— the improvement of the means of signal communications.[1]

---

[1] The original French text uses the word "*transmissions*" for signal communications involving the transmission of messages between units and the word "*communications*" for the network of land, water, and air routes that connect forward military units to their bases of operation used for the movement of troops, matériel, and supplies. To avoid confusion, these two terms have been translated throughout as "signal communications" and "communication routes," respectively.—Trans.

Fortified fronts.

Built to protect the national territory from invasions, the new permanent fortification *ouvrages*[1] now make it possible:

— to mobilize under the protection of a solid covering force that is relatively economical in manpower:

— to protect our major industrial regions and the sensitive points of our frontiers as widely as possible;

— to provide a powerfully equipped base of maneuver for our armies.

The tactical role of the fortified fronts, their organization, their occupation, and the methods of their defense are defined by this Instruction.

They are dominated by the irrefutable lessons of war: *the permanent fortification must be merged and integrated into the general disposition of the armies.*

Mechanization and motorization. — Antitank weapons.

However powerful the fortified fronts may be, the decision, the same tomorrow as yesterday, will only be obtained by maneuver, the essential elements of which are speed and mobility.

Advances in motorization make it possible to move large forces quickly; advances in mechanization make it possible, through the creation of appropriate units, to guarantee the safe movement of these forces and to support their action.

Thus, new forces appeared alongside the large units of the standard type, assembled in modern large units and forming a "system" capable of carrying out the battle with its own means.

In this system, *the Cavalry Division has given way to the Light Mechanized Division, and the Infantry Division has been transformed into a Motorized Division.*

Such Large Units (D.L.M.[2] and D.I.M.[3]) obey, like their predecessors, the immutable principles that govern the employment of the armed forces.

---

[1] *Ouvrage* = (n) literally: works; an autonomous element of a fortified organization, able to resist even after encirclement. These were the major fortification elements of the Maginot Line.—Trans.

[2] D.L.M. = Light Mechanized Division (*Division légère mécanique*). Despite its name, the D.L.M. was a true armored division and was organizationally very similar to the German Panzer Division. They were labeled "light" divisions only because the more impressive title of *Division mécanique* would have caused a confusion of acronyms with the Moroccan Division (*Division marocaine*).

The French organized its first "Armored" Division (*Division cuirassée*) as part of the infantry arm in January 1940. Initially called Heavy Mechanized Divisions (*Division lourdes mécaniques*) and based on heavy mass-maneuver tanks, they were renamed to highlight their differences from the cavalry's Light Mechanized Divisions. In 1943, the French army adopted the name *Division blindée* for the multi-purpose armored divisions it organized in North Africa with U.S. assistance. This latter term has been retained for all subsequent French armored formations.—Trans.

[3] D.I.M. = Motorized Infantry Division (*Division d'infanterie motorisée*).—Trans.

But, of these permanent principles, that of security, in particular, must be, if not revised, at least enhanced to continue to protect, on the scale of the motorized system, the commander and the troops against surprise, made more dangerous by the existence of modern, numerous, and rapid means on the part of the potential adversary.

Parallel to the creation of mechanized or motorized large units, the improvement of armored vehicles, particularly tanks, has continued unabated.

Their speed increased considerably while the thickness of their armor continued to progress concurrently with advances in antitank armament.

From this bitter and new phase of the age-old struggle of cannon against armor, vigorously waged over the last few years and whose evolution is doubtless not finished, certain lessons had to be learned. This Instruction has, therefore, endeavored to identify, in the current state of armaments:

— the conditions for the use of the tanks;

— the methods of defense against enemy armored vehicles.

Regarding the use of tanks, it cannot be emphasized enough that today, antitank weapons stand before the tank as, during the last war, the machine guns did before the infantry.

The Commission, taking into consideration this formidable danger, was of the opinion that the considerable development of the number and power of antitank weapons in all foreign armies has led to planning for the use of tanks in the attack only with the protection and support of very powerful artillery. It only considered their deep action after prior disorganization of the adversary's defensive system, particularly in the exploitation of success where armored vehicles seem to have to obtain decisive results.

As for the defensive measures to be taken against enemy armored vehicles of all kinds, the new Instruction highlights the essential interest in screening the dispositions with natural obstacles, in the first place by rivers, the crossing of which will always pose a difficult problem even if the use of amphibious vehicles were to become generalized.

It also underlines the importance of the *in-depth organization* of defense against tanks, wanting fire combined with obstacles.

Air forces and air defense forces.

Advances in aviation, marked by a considerable increase in aircraft power, speed, and range, offer the armies greater possibilities for cooperation. Intelligence is sought at very long distances, target identification and the placement of fire can be carried out at the greatest ranges of modern artillery, and heavy aviation finally allows the command to make its action vigorously felt throughout the depths of the battlefield.

On the other hand, the parallel development of foreign aviation requires a corresponding organization and reinforcement of the various elements of air defense (defensive light aviation, antiaircraft defense, territorial antiaircraft defense) to allow the dispositions of the land forces to evade the investigations of the opposing air force and repel its attacks.

The new Instruction devotes the necessary developments to aviation's action in battle and to the combination of antiaircraft defense forces. It has also endeavored to draw attention to all the progress made in the air domain, whether it is a question of using them or avoiding their effects (autogyro or airdrops (*descente aérienne*)).

Signal communications.

The leading role played by signal communications in modern armies has caused the Commission to define general rules for its use within the framework of the maneuver of large units.

In this connection, it is important to underline the continuous progress of radio processes, which, subject to a suitably disciplined use, render the maneuver more flexible and the combination of arms closer and safer.

The new means we have just reviewed have further developed the firepower that the drafters of the 1921 Instruction had already described as overwhelming. This firepower will be exercised tomorrow on the battlefield, where it will reign supreme, with *increased violence* and *depth* due to advances in bomber aviation and the longer ranges of modern artillery.

The value of defensive organizations has seen a parallel development, and this synchronism of the progress made in the fields of fire and protection has had the effect of preserving in operations the fundamental traits of their appearance and of keeping to the different Large Units their respective tasks in the battle.

In the offensive, the embrace of maneuver develops gradually. The adversary is first tackled flexibly on all fronts (establishing contact); he is then probed at numerous points to verify his effective presence or to assess his strength (engagement); and finally, the attack aims to deal him a fatal blow.

Each Large Unit has a specific role in this maneuver:

The *establishment of contact*, made by the cavalry, is clarified by the advance guards. Their actions are coordinated by the generals commanding the first-echelon divisions.

The *engagement* includes local actions that are often combined, requiring only a small amount of manpower but energetically supported. It is carried out with the means of the first-echelon Large Units. It is controlled in principle by the generals commanding the first-echelon *Army Corps* within the framework of the instructions received.

The *attack* is the work of the army commander, applying the maximum means, strongly centralized in his hands, in the desired direction.

Advances in armament and the generalized use of mechanized means have a more marked influence on the features of the *defensive* and the general missions of the Large Units taking part in it.

Far from weakening the scope of the rules established in 1921 concerning the occupation of a defensive position, which are based on the echelonment in depth of means that combine their fire in front of a main line of resistance supported by an obstacle, these new data tend, on the other hand, to increase the advantage of the *obstacle* and the importance of *echelonment in depth.*

The obstacle, preferably natural, must be sought first by all the elements responsible for the defensive mission; its presence most often determines, today, the choice of the position of resistance for which it provides immediate and effective protection against the actions of armored vehicles.

This same threat, together with the increased range of artillery, leads, on the other hand, to the echelonment of the means of defense over a greater depth in order to disperse the risks, to impose on the enemy successive efforts and to give the defenders, at any point in the position, a base for resistance in place or for counterattack.

The echelonment in depth assumes an extent that varies with the importance of the Large Unit concerned. More specifically, it is the responsibility of:

— the *Division* and the *Army Corps,* in order to hold a defensive position without thought of retreat (*sans esprit de recul*)[1], to echelon their means throughout the depth of this position provided that they remain in a position to carry out a concentration of fires as dense as possible in front of the main line of resistance;

— the *Army* to give the defensive battle even more depth and flexibility by ensuring, behind the army's position of resistance, the possible defense of successive positions well protected against the penetration of armored vehicles.

---

[1] *Défense sans esprit de recul* was a central part of French military doctrine, but this term is difficult to translate precisely into U.S. military terminology. French training manuals defined it as "holding a given position in spite of the enemy." The closest U.S. military term is "defense in place", which is defined as "a system of defense based upon firm resistance without retreat, as opposed to delaying actions in successive positions." U.S. military dictionaries of World War II have also translated it as "obstinate defense," "defense at all costs," and "sustained defense," each of which has a subtly different meaning. A confidential 1939 report by the U.S. Army's Infantry Board on the doctrine of the French Army stated that the defensive doctrine "does not contemplate an elastic or flexible defense for division or lower units. Every unit, and every man, is trained to remain in position to which ordered, unless they receive positive orders to the contrary. . . . Their infantry is taught that it must not necessarily consider the situation hopeless if their individual center of resistance is surrounded. Further, if they are surrounded, by hold, it will be a matter of time until higher command will, using additional troops, drive the enemy back and, thus re-establish the original position."—Trans.

ADDITIONS.

After having endeavored to adapt, as it has just been said, the tactical processes to the progress of sciences applied to the armament and transportation of the troops, the new Instruction brings to that of 1921 the following additions:

I. The distancing in time from the previous war, the regular renewal of army officers, and the commissioning of new resources prompted the high command to ask the Commission to present *a tableau of the offensive and defensive battle,* a kind of compendium of the general principles of the conduct of large units and the combination of arms. This Title, entitled *The Battle (Title V),* constitutes the necessary preface of the following Titles (VI, VII, and VIII), which treat the use of the Army, the Army Corps, and the Division in battle;

II. The Instruction of 1921 devoted very little space to the *cavalry.* It was silent, in particular, on the conditions under which the command could employ the Large Cavalry Units.

This gap has been filled by the introduction in this Instruction of Title IX entitled *General points on the use of Large Cavalry Units;*

III. It appeared advantageous to disseminate widely *the analytical method according to which the problems of war are reasoned,* such as it has long been identified and taught in the higher schools of military studies;

IV. In order to limit the terminology to be used in wartime operations and to identify the meaning of certain terms already consecrated by usage or made necessary by new armaments, a small number of *essential definitions* have been retained. They appear at the beginning of the Instruction.

Similarly, the Commission deemed it useful to modify certain designations.

1) It decided, in particular, to put an end to the confusion caused by the similarity of the terms: *direct-support grouping* (*groupement d'appui* direct) and *direct-support fire* (*tir d'appui direct*).

The first of these designations, which fortunately expresses the mission (support) of an artillery grouping working without intermediary (direct) with such an infantry or tank unit, has been retained. On the other hand, the expression *direct-support fire* has been replaced with *close-support fire* (*tir de soutien immédiat*).

This modification will have the advantage of no longer letting people believe that a direct-support grouping is only called upon to provide "direct support" fire when it can actually execute any fire requested by the unit it is supporting or by the command.

2) Regarding tanks, it seemed preferable to the Commission to renounce tightly holding them in a rigid classification based either on weight or on a mission definitively assigned to a given tank.

It considered, in fact, on the one hand, that given the progress of industrial technology, the distinction of tanks into light, medium, and heavy tanks no longer gave precise indications of the possibilities of these machines; it admitted, on the other hand, that the majority of the tanks were able to fulfill all the missions, from accompanying the infantry to deep penetration into the opposing disposition.

Under these conditions, the Commission resolved to designate the tanks by *their current mission* and, in particular, to call by convention, whatever their model:

— *accompanying tanks* (*chars d'accompagnement*), tanks subordinated to the infantry and having received the mission to accompany the latter in combat;

— *mass-maneuver tanks* (*chars de manœuvre d'ensemble*), tanks kept under the orders of the commanders of the Large Unit and charged with fulfilling a particular mission for the benefit of the maneuver of the Large Unit.

This Instruction will be supplemented later by the annexes below:

Annex 1. — Instruction on field service;

Annex 2. — Instruction on liaison and signal communications;

Annex 3. — Instruction on intelligence and observation;

Annex 4. — Instruction on gas protection;

Annex 5. — Instructions on the organization of the terrain, camouflage and destruction;

Annex 6. — Instruction on communication routes and transportation;

Annex 7. — Instruction on mountain operations.

All of these regulations will also be accompanied by a notice intended for the high command only which will aim to outline the strategic framework in which the Large Units would be called upon to be employed.

COMPOSITION OF THE COMMISSION.

General GEORGES, member of the Higher War Council.

General PUJO, Chief of General Staff of the Air Force.

General MEULLÉ-DESJARDINS, Commander of the 5th Region.

General DUFFOUR, Commander of the 30th Region.

General BOUCCHERIE, former Commander of the 1st Cavalry Division.

General HUBERT, Commander of the Metz Fortified Region.

General LOIZEAU, Commander of the 12th Infantry Division.

General TOUCHON, chief of the Infantry Tactical Studies Section of the Combined Arms Tactical Studies Center.

General MONTAGNE, Commander of the 258th Infantry Division.

General SISTERON, General Staff of the Higher War Council.

General SIVOT, Commander of the School of Engineering.

Colonel MENDRAS, Commander of the 182nd Heavy Artillery Regiment.

Colonel PUGENS, Commander of the 188th Infantry Regiment.

Colonel KERGOAT, chief of the 4th Bureau of the Army's General Staff.

Lieutenant-Colonel TARRADE, professor at the Liaison and Signal Communications School.

Major NAVEREAU, General Staff of the Higher War Council.

GENERAL DEFINITIONS

Large Unit. — The assembly in an organically constituted framework, under the command of the same leader, of troops of different arms and the services necessary for them to live and to fight.[1]

Motorized unit. — Unit organically equipped, in whole or in part, with motor vehicles of the current type, wheeled or tracked, which ensure its transport but do not essentially modify the tactical mode of use of the unit in combat.

Mechanized unit. — Motorized units based on armored vehicles.

Armored vehicle. — Motor vehicle, in principle all-terrain, carrying weapons under armor and able to fire on the move.

Concentration. — Operation aimed at bringing together, in the border region, the forces forming part of the initial disposition prescribed by the Commander-in-Chief for the start of a war.

This term is also used to designate an assembling of forces performed during a campaign for an important operation.

Base of concentration is the zone in which the concentration takes place.

The set of arrangements made for the grouping of forces on the base of concentration constitutes the *concentration plan.*

All of the measures provided for transporting the units to the base of concentration, *known as concentration transport,* constitute the *concentration transport plan.*

The elements responsible for ensuring the protection of a concentration constitute the *covering force*: the distribution, missions, and methods of action of the covering troops are contained in the *covering plan.*

Air cover protects the concentration against enemy air incursions.

Direction. — Every mass maneuver (*manœuvre d'ensemble*) of a Large Unit includes a direction on which the commander must

---

[1] In practice, Large Units included divisions, army corps, armies, army detachments, and army groups.—Trans.

strive to maintain or restore the center of gravity of his disposition during the maneuver.

Depending on this direction assigned to the Large Unit, each subordinate unit receives a direction of its own.

Axis of effort. — Axis temporarily chosen by a Large Unit to overcome a specific obstacle. This should not be confused with the direction, but it must ultimately contribute to leading the Unit in this direction.

Objective. — Strip of terrain whose conquest will ensure the success of the Large Unit's maneuver.

If the objective cannot be achieved in one push, then the operation is divided into phases. Each of these phases corresponds to a normal objective and usually a possible objective.

Normal objective — assigned to a Large Unit. — The objective that it must strive to reach with the main body of its strength by committing all its means if necessary.

Possible objective. — An objective that the Large Unit can aim at beyond the normal objective if circumstances become favorable.

Intermediate objective. — Within a phase, a subordinate Large Unit may choose or receive intermediate objectives.

Bounds. — Successive stages in the methodical advance of a unit.

Zone of action. — Extent of terrain on which a unit fighting abreast of two others is called upon to deploy and use its means. It corresponds, in the offensive, to the front on which contact will be sought and, in the defensive, to the front occupied by the first-line units.

Disposition. — All the means of a unit articulated for the execution of the mission and in accordance with the commander's idea of maneuver.

Tactical grouping. — Temporary assembly of elements of different arms (or detachments of arms) under the command of the same leader, within the framework of a Large Unit, for executing a specified tactical mission.

Contact front. — Apparent outline of all the contacts made over the entire width of the unit's zone of action.

Engagement front. — The one on which the Large Unit commits part of its means to clarify or complete the results obtained by the establishment of contact.

Attack front. — The one on which the unit is likely to carry out a force action with the organic and reinforcement means at its disposal.

Stabilized front. — Front that received, during the operations, a comprehensive defensive organization, exploiting all the resources of field fortifications and matériel.

Fortified front. — Defensive front supported by permanent fortification works.

TITLE ONE.

THE COMMAND AND COMMAND ORGANIZATION.

CHAPTER ONE.

THE COMMAND.

ARTICLE 1.

THE LEADER. — HIS ROLE.

1. The leader's *personality* exerts a major influence on the design and conduct of wartime operations.

Judgment, determination, character, and a *taste for responsibility* are its essential traits and dominate all the physical, intellectual, psychological, and technical qualities that any commander of a Large Unit must possess.

The sense of *duty* and the practice of intellectual *discipline* must, however, maintain his action within the limits defined by the mission entrusted to him by higher authority.

2. The commander of a Large Unit is careful not to enclose his subordinates in narrow prescriptions likely to restrict the execution of their invigorating initiative. In addition, he must *anticipate:* his forecasts are broader and more distant as he occupies a higher position in the hierarchy.

Possessing to a high degree the *sense of tactical possibilities,* he must also take into account, at all times, the matériel and psychological state of his units, the good condition of which often permits all daring.

For this, it is necessary that he knows his troops and makes himself known to them, that he sees them as often as possible, watches over their well-being, and endeavors to spare them unnecessary fatigue.

His energetic and calm attitude, his firmness, and his benevolence, complete to win him the indispensable *confidence of his subordinates.*

3. The primary role of a Large Unit Commander is to *design, plan,* and *conduct* Large Unit operations.

ARTICLE 2.

GUIDING PRINCIPLES.

4. The leader conceives, prepares, and conducts his maneuver based on the guiding principles, which, by their character of generality and permanence, are the basis of all wartime operations.

5. These principles are:

— ***to impose his will on the enemy,*** by the action of the main body of his forces, offensively or defensively, under the most favorable conditions of place and time;

— ***maintain freedom of action,*** despite the enemy's actions;

— distribute his forces between the various missions according to the rules of a ***strict economy***[1], by devoting the maximum of means to the primary mission and by endeavoring to obtain the best return, on the one hand, from each of the elements, on the other hand of their combination.

The application of these three principles constitutes the very essence of the maneuver: to maneuver is to conveniently achieve concentrations of force and combinations of efforts in spaces or directions corresponding to the intended goals.

The leader imposes his will on the enemy by endeavoring to carry out thoroughly and without any intention of reversing the decisions taken for attack or defense; he adapts the general arrangements of his plan to events, but he leads his maneuver to the goal he has fixed for himself.

Maintaining freedom of action allows the leader not to submit to the enemy's will and to engage the main body of his forces only when and where he chooses. It is obtained through *intelligence* and the organization of *security.*

If the enemy forestalls him, he opposes him with only a part of his forces as small as possible in order to conserve for himself powerful means to ensure the execution of his own projects.

The combination of economy of forces and security leads to the *deployment* of these forces.

Both strategy and tactics result in a series of successive deployments, making it possible to produce the *dispositions* corresponding to the consecutive problems to be solved.

---

[1] The economy of forces must be understood here in the sense of a fair distribution of them.

6. Successes in war are achieved:

— by striving to achieve *surprise* in maneuver, in the attack as well as in the defense;

— by making *long-term forecasts* so as not to be surprised by the development of events.

To any maneuver revealed, the adversary can respond with a countermaneuver; the leader must, therefore, strive to surprise the enemy. *Surprise* is obtained as much from the secrecy in the preparation and *speed* in the execution as from the novelty or the unexpectedness of the means and processes.

The need for long-term forecasts is all the more important to the leader when the masses to be implemented are greater and more mobile, and when his zone of action is more extensive.

He must consider all eventualities in advance, measure the consequences, however remote they may be, of the enemy's possible initiatives as well as those of his own maneuvers, and reflect on the action to be taken in each case.[1]

It is necessary, indeed, to focus on war as much as possible, to suppress or, in any case, to correct and to use chance.

These rules do not exclude bold solutions, which are sometimes even the wisest or remain the only way out, provided that they are executed methodically.

ARTICLE 3.

THE DECISION.

7. The commander of a Large Unit, after having examined and put in their exact plan all the conditions of the action to be undertaken, matures the conception of his maneuver, and translates it into a *decision,* simple, clear, and rigorously in conformity with the intended purpose and likely to be achieved in the minimum of time, with the fewest losses and the best results.

I. ELEMENTS OF THE DECISION.

8. The main elements to be considered by the leader, called upon to decide on the tactical employment of a unit, are:

— the *task,*

— *the means and time* at his disposal,

— *the terrain,*

---

[1] To this end, in the methods of command and instruction, while perfecting the reflexes, one must develop the habit, the higher one rises in the hierarchy, of reasoning quickly but carefully about the problems before making one's decision.

— *the strength and capabilities of the enemy.*

The analysis of these elements, then their synthesis, constitutes the method of reasoning for wartime operations.

It is essential to undertake them in a correct mood.

A wartime action, in fact, is never isolated in time or space; the leader must, therefore, before any other study, carefully consider the very diverse conditions which together create the *general situation* in which the envisaged operation takes place.

He will then be in a position to carry out a valuable examination of the factors which are the subject of the following paragraphs:

The mission.

9. The mission must clearly define the goal to be achieved. It is given in a clear form that does not lend itself to any interpretation or ambiguity. It engages the responsibility of the leader who gives it. It is imperative for the one who receives it.

The mission must be constantly in the mind of the leader in the design as well as in the execution.

All the arrangements he makes tend to accomplish it by conforming fully and constantly to his spirit.

He strives to pursue its execution without deviation until the end.

The means.

10. The leader must have an exact knowledge of the means in personnel and matériel put at his disposal for his operation.

From their possibilities, judiciously evaluated, he will deduce the scale to be given to his maneuver, taking into account the time at his disposal.

In war, the value of a troop is measured by its ability to attack and to resist.

But the power of the fire and the intensity of the combat create rapid wear on the most seasoned troops. It is, therefore, important to provide the units with the essential rests to catch their breath, to reconstitute, and to resupply themselves.

In all his combinations, the leader must consider the probable repercussion of the events on the morale of the troop.

In this respect, the first engagements of a war are of particular importance.

The value and cohesion of the young troops participating in it must be predicted cautiously and lead the command to provide them with particularly energetic matériel support.

It is always best to keep units under their usual leaders; the severance of organic ties can only be an expedient temporarily imposed by circumstances.

The perfect knowledge of the constraints and the performance of the matériel, particularly armament, signal communications, and transport matériel, is at the base of any coordinating action.

It is also essential for the commander to know the possibilities of new matériel, particularly those that have not yet appeared on the battlefields. He will often find in their use a surprise factor and an element of undisputed superiority over an adversary who lacks it.

The terrain.

11. Terrain exercises a tyrannical influence on any maneuver.

As part of the planned operation and on the scale of the personnel to be implemented, the terrain is studied, first on the map, then by ground or aerial observation, to make full use of its properties, to respect its constraints and to discriminate, in an objective way, the possibilities which it offers to the maneuver of the enemy.

12. In an Army Group or Army operation, the major forms of the terrain, its nature, and the network of communication routes must be taken into particular consideration.

Significant irregularities of the ground (mountainous massifs, large terrain breaks (*coupures*)[1]), as well as extensive forests, constitute obstacles to the movement and deployment of the Large Units.

The areas of easy routes, particularly the thresholds between two rugged regions or open valleys, often present invasion routes or routes to be used or denied.

The nature of the terrain, open or closed, free or broken, flat or hilly, must also, under pain of serious miscalculations, hold the attention of the commander responsible for designing a large-scale maneuver. Closed, broken, or uneven terrain places a heavy burden on the engagement of Large Units and the combination of their maneuvers.

The abundance of railroad and road communication routes available to the two adversaries gives the maneuver both power and speed.

Their rarity, by paralyzing transport and supplies, compromises or restricts any important action.

Careful study of the terrain will enable the leader to determine the regions where he can advantageously accept battle or force the enemy to receive it.

It will lead him to choose successive and contingent *battlefields*, the prior determination of which will subsequently dominate the conduct of his maneuver.

These battlefields should, as far as possible, present:

— toward the enemy, good clearances for observation and fire, and for the cooperation of the various arms;

---

[1] *Coupure* = (n) cut; in the military context, a geographical terrain feature opposing the advance of ground troops (rivers, streams, deep valleys, etc.).—Trans.

— to the rear, communication routes and cover suitable for a powerful, rapid, and safe inflow of reinforcements and matériel.

13. In a Corps or Divisional operation, surveying the terrain is undertaken from a primarily combined arms viewpoint.

The ground relief and the nature of the terrain are of particular interest in this respect.

Terrain that is generally open, undulating, easy to navigate, without major natural obstacles, lends itself, by the extent and continuity of views it allows, to the combination of arms and the reciprocal support of units in combat. The motorized or mechanized units of the two adversaries find fruitful use there.

Fire there is effective, movement is rapid, deployment easy, and the exercise of command is facilitated. However, great bare plains are poorly suited to offensive action by infantry and tanks, at least in mobile warfare, due to the extent of the fields of fire of an installed defense and the increased power of aviation.

A very covered, broken, uneven ground makes movement difficult. It requires, for the infantry, more manpower, but it facilitates elementary maneuvers. Obstacles and cover limit weapon effectiveness; the combination of infantry and artillery fires, and the exercise of command are precarious because of the difficulties of observation.

The deployment of means is difficult, although their concealment is assured. The action of armored vehicles is diminished or eliminated by obstacles, and the localization of their use attenuates their psychological effect.

The study of the terrain, from the tactical point of view, notably leads the commander:

— *in the defensive,* to choose the strip of terrain on which he intends to place his fire barrage;

— *in the offensive,* to determine the directions and the objectives, as well as the compartments of terrain in which the combined action of the arms will be able to develop favorably, sheltered from the sights and the distant fires of the adversary.

Fortifications.

14. The value of terrain is significantly increased by its organization.

The purpose of this is to increase the importance of obstacles and covers or to create them at judiciously chosen points; it thus makes it possible to install, sheltered from destruction by the enemy, the means of fire, and the personnel in charge of implementing them.

The organization of the terrain can be either permanent or carried out at the time of need and according to the necessities of the maneuver. In the first case, it takes the name of *permanent fortification,* and in the second, that of *field fortification.*

15. From a strategic point of view, permanent fortifications provide the high command with important advantages:

— they give it the possibility of holding large sections of the theater of operations by practicing a rigorous economy of forces; it facilitates, in particular at the beginning of a conflict, the mission of protecting the mobilization (*couverture*);

— they allow it, sheltered by the fortified front guaranteed against intervention by the enemy's ground forces, to carry out large concentrations of forces, to prepare his maneuvers and, after having completed his musters, to debouch[1] for an attack;

— finally, they can contribute to screening its communication routes or one of its flanks in an offensive or defensive maneuver, thus making the fortified zone a strong point or a pivot point for the maneuver.

From the *tactical point of view,* the permanent fortifications provide the means to hold certain parts of the terrain very solidly with relatively small numbers and to obtain powerful effects from the fire by carefully adjusted fires from fixed and protected matériel with high rates of fire.

Conversely, it obliges the enemy who wants to seize permanent fortification *ouvrages* to take actions of extreme violence, very costly, generally very long, often absorbing a significant part of his forces, and always requiring the implementation of numerous and powerful matériels.

16. *Field fortifications* have advantages similar to permanent fortifications, but to an extent related to the development given to its organizations.

It can often acquire considerable value when combined with destruction and with the use of significant natural obstacles.

Flexible and progressive, it must be used in all circumstances of the combat, in particular by any troop halted, however short the duration of its halt.

The enemy.

17. The leader endeavors to make ***a complete examination of the possibilities of the enemy.*** To this end, he uses:

---

[1] *Debouch (debouchment).* To emerge from cover into an open area under enemy fire or ground observation. For example, "The rifle elements debouched from the final assembly areas following the rear elements of the attacking light tanks."

b. To emerge from a defile into a wider, more open area. For example, "The battalion debouched from the eastern exits of the village." (Definition from U.S. War Department *FM 7-5: Infantry Field Manual: Organization and Tactics of the Infantry, The Rifle Battalion,* 1940).—Trans.

— on the one hand, documentation gathered up to now on the general value of the adversary (armament, methods of combat, lines of communication networks, etc.);

— on the other hand, information which is more particularly important for him to know for the operation envisaged and which he has taken care to have sought, taking as a basis his own plans (No. 121).

He thus examines all the enemy's possibilities of maneuver assessed in their order of priority and probability; because of his powerlessness to know the real intentions of the adversary, moreover variable during execution, a leader can only hope to lead his maneuver to success if he is constantly ready by his disposition to face the main contingencies.

II. EXPRESSION OF THE DECISION.

18. The conception of the leader translates into a decision that manifests his will to all.

This power of decision confers on the command a responsibility that belongs only to him and which constitutes one of his highest prerogatives.

The decision must clearly indicate the form and the essential characteristics of the maneuver designed by the leader. It specifies to each of his immediate subordinates his mission and his means.

It must be taken in a timely manner, taking into account the time necessary for the transmission and execution of orders.

Documentation on the enemy will always be incomplete: to wait to be fully informed to act is to risk inaction.

The plans.

PLAN OF MANEUVER.

19. The leader of a large unit expresses his decision in a personal document in which he clearly outlines his *idea of maneuver* and specifies the essential arrangements of the operation he has designed.

It is the *plan of maneuver,* which can be called a *plan of attack* or a *plan of defense,* depending on whether it is an exclusively offensive or defensive operation.

20. The higher commander's plan of maneuver specifies for all Large Units:

— the general aim pursued,

— the strategic attitude adopted accordingly,

— the idea of maneuver,

— the direction of the maneuver,

— the objectives to be achieved or the positions to be defended,

— the initial disposition of the forces.

Within this framework, he then determines the missions of the subordinate Large Units.

21. The plan of maneuver of the commander of a Large Unit also contains:

— the idea of maneuver,

— the initial disposition,

— the missions of subordinate units,

— their directions and zones of action,

— the objectives to be conquered or the positions to be defended,

— the general conditions for carrying out the maneuver.

22. The subordinate leader reports on his decision to the higher authority which assigned him his mission, either by communicating to him his plan of maneuver himself, or by simply sending him, as a report, the summary of this plan.

INTELLIGENCE PLAN.

23. Knowledge of the enemy being one of the main elements of the commander's decision, he has the duty to establish, according to his mission, the methodical inventory of the intelligence to be collected before and during any operation envisaged,

He thus specifies the points that it is useful for him to know to conduct his maneuver, avoid surprises, and increase the chances of success.

These data constitute the *intelligence plan.* This plan, a function first of all of the initial situation, changes during operations as a result of events, and due to the evolution of the maneuver; it must be constantly updated (No. 123).

LIAISON PLAN.

24. Because of the vital importance of the rapid transmission of orders, intelligence, and reports, a *liaison plan* is drawn up for each operation. This plan serves as the basis for the signal communications maneuver, which must be closely adapted to the plan of maneuver of the command.

SERVICE EMPLOYMENT PLAN.

25. The importance of supplies and evacuations in combat, the considerable influx of matériel of all kinds required by the battle, oblige the command to regulate, in advance and with care, the deployment of resources, the distribution facilities, and the use of means of transport. In addition, because of the constant movements occasioned by the movements of the troops and the interplay of the services, the maintenance and operation of the lines of communication must be the subject of precise prescriptions.

The decisions taken on this subject are compiled in a *service employment plan*, the data of which is sent to the services in good time.

The service employment plan must cover not only the initial organization but also the modifications it is called upon to undergo according to all forecasts.

Instructions and orders.

26. The leader communicates his decisions to his subordinates through instructions and orders. At the level of the commander-in-chief, the instructions frequently take the name of directives.

The purpose of the directive or *instruction* is to guide the immediately subordinate authorities to enable them to act in all circumstances in accordance with the views of the command. To this end, it makes known the commander's intentions, the goal, the general idea, and the rhythm of the maneuver; it sets for each Large Unit, its mission, its direction, and its objectives; it provides for the different eventualities and the resulting methods of action.

It is established for a more or less long period according to the level of command from which it emanates and the nature of the planned operation, while avoiding exposing too distant forecasts that events could deny or render obsolete.

It most often has a *personal character*; its prescriptions must be kept *secret* by the addressee and only be notified to the staff or subordinate authorities to the extent strictly indispensable and never by the communication of the document itself.

The order contains precise, imperative prescriptions, generally applicable in the short term and under clearly determined conditions.

The *order* is *general* or *particular*, depending on whether it is addressed to all the elements of the Large Unit or only to some of them.

The *warning order* is intended to give the troops the necessary indications so that they can make their initial arrangements in good time. It is a constant practice in mobile warfare operations.

Orders and Instructions ***should be brief,*** especially in mobile warfare. They must, however, contain all that is necessary to make clear the intentions of the command.

CHAPTER II.

COMMAND ORGANIZATION.

ARTICLE 1.

COMMAND OF LARGE UNITS.

27. In principle, Large Units are commanded by general officers of the rank of Major General (*général de division*), but whose rank and prerogatives vary according to the command they exercise.

When the Commander of a Large Unit is absent, he is temporarily replaced in his functions by the senior officer of this Large Unit with the highest rank: however, another officer can be designated in advance to, if necessary, exercise this command; he is then provided with a letter of appointment.

The Commander of a Large Unit has a staff.

The staff.

28. Through its work and its reconnaissances, the staff prepares the commander's decisions.

It is in charge of drafting orders intended for the troops and the services, monitors their execution, and provides the necessary liaison to inform the command of the situation, the needs of the troops, and the functioning of the services.

The high mission of the staffs requires officers who have superior qualities of general and professional education, tact, devotion, and self-sacrifice.

The staff is the auxiliary of the command; it has no authority of its own over the troops or the services; in a Large Unit, the responsible implementation bodies are the commanders of the subordinate Large Units, the commanders of the troops, and the chiefs of service.

In each staff, the whole of the service is directed by a *Chief of Staff* who, in the Large Units above the division, is assisted by a Deputy Chief.

The Chief of Staff is the immediate collaborator of his general; confidant of his intentions, he is able to foresee and prepare in good time the elements of his decisions, and then the resulting execution measures.

He is qualified to sign by order; he may, under his responsibility, partially delegate his right to signature to certain staff officers.

The Chief of Staff is responsible, by delegation of command, for directing the action of the various services.

He directs the work of the staff and pays special attention to the study and interpretation of intelligence; he particularly monitors and controls issues relating to the organization of liaisons and signal communications and oversees their operation within the Large Unit.

ARTICLE 2.

COMMAND OF ARMS AND DIRECTION OF SERVICES.

29. To facilitate the exercise of his command, the commander of a Large Unit generally keeps with him, in addition to his staff, the chiefs of certain arms, as well as the directors or chiefs of services.

The meeting of the staff, the personnel of the command of the arms, and the direction of the services in a Large Unit constitutes the headquarters of this Large Unit.

TITLE II.

MEANS AND MODES OF ACTION.

CHAPTER ONE.

MEANS OF ACTION.

ARTICLE 1.

GENERAL.

30. Wartime operations between Great Powers may embrace several regions, distant or separated from each other. Each of these regions constitutes a *theater of operations.*

The forces acting in a theater of operations are under the orders of a Commander-in-Chief, of the rank of Marshal of France or Major General, who reports directly either to the Government or to a Commander-in-Chief of the all the French armies spread over several theaters of operations, or possibly an allied commander-in-chief.

His letter of command specifies, if necessary, his allocation of air and naval forces acting in the theater of operations.

31. Troops of all arms have the task of leading the combat, closely combining their action and coordinating their efforts in accordance with command orders.

The various arms are:

— infantry, including tanks;

— artillery;

— cavalry (horse, motorized, mechanized, or mixed units);

— engineers, including signal communications troops;

— air forces and antiaircraft defense;

— the train.

The services are responsible for meeting the needs of the command and the troops in all circumstances.

32. In the field, troops and services are generally grouped into Large Units: ***Large field units and Fortress Large Units.***

The Large field units are:

— *the Infantry Division;*

— *the Cavalry Division;*

— *the Army Corps;*

— *the Cavalry Corps;*

— *the Army.*

With a view to special and generally temporary missions, elements of troops and services of varying size may be brought together in an *Army Detachment.*

When the number and the importance of the armies of the same theater of operations justify it, *Army Groups* can be formed.

Each brings together, under the same command, subordinate to the commander-in-chief, a variable number of armies, divisions, or cavalry corps.

The Fortress Large Units are:

— *the Fortified Region Mixed Brigade;*

— *the Fortified Region.*

Finally, other elements (troops or services) constitute *General Reserves* at the disposal of the higher command. These elements, as soon as they have been assigned to a Large Unit or a detachment, are considered to be an integral part of it from the point of view of their use.

33. Field and Fortress Large Units, General Reserves, and, when their importance justifies it, detachments, include as constituent elements:

— command and headquarters;

— troops;

— services.

ARTICLE 2.

CHARACTERISTICS AND GENERAL ORGANIZATION OF ARMS

I. Arms.

Infantry and tanks.

34. ***Infantry is tasked with the primary mission in combat.***

Protected and accompanied by its own fire and by artillery fire, possibly preceded and supported by tanks, aviation, etc., it conquers the terrain, occupies it, organizes it, and preserves it.

Its task is particularly harsh, but glorious above all.

In combat, it is the infantry that exerts itself the most; all the efforts of the command must, therefore, tend to spare it, to maintain or to exalt its morale and its energy.

35. Infantry acts by fire and by movement.

Infantry fires are characterized by their precision, diversity, and power at short and medium distances.

They can also exercise delaying and interdiction actions up to the maximum distances allowed by the methodical execution of the fires.

The infantry is equipped with:

— weapons with flat-trajectory fire, generally automatic, and with high rates of fire, whose grazing actions are particularly effective;

— curved-fire weapons, of small and medium caliber, which enable it to attack an enemy who is sheltered or concealed;

— antitank weapons and antiaircraft weapons.

— edged weapons (*armes blanches*).[1]

We distinguish:

— *light* weapons, carried by a single man;

— *heavy* weapons, broken down for transport into several loads or towed.

In the *defensive*, its armament gives the infantry considerable stopping power, even on extended fronts. This power takes on its full value by organizing the terrain as thoroughly as possible.

36. *Fortress infantry* are organized to fight:

— either in the *ouvrages*;

— or in their intervals.

---

[1] Literally "white weapons": cutting, thrusting, or blunt weapons used in close combat (such as a knife, bayonet, or club) as distinguished from a firearm.—Trans.

It is equipped with an organic armament and serves the arms, matériels, and special weapons with the various fortification units.

The troops that are part of the garrison of an *ouvrage* have an organization corresponding to the importance and the nature of the armament and the matériels which they deploy and the needs for the own defense of the work.

The perfect knowledge of its mission, the terrain, and its means, the meticulous preparation of fire plans, the use of reliable observation and signal communications networks, in addition to the certain protection of its primary means of action, allow fortress infantry, supported by artillery, to stop enemy attacks through the use of a precise, adjusted, and powerful fire system.

37. *Tanks* are armored vehicles capable of moving more or less quickly, depending on their technical qualities.

The difficulties of the terrain, natural or artificial, can significantly reduce their speed and mobility.

They *are able* to perform passages in certain passive obstacles and to neutralize or destroy, up close, active resistances.

They can, on their own, in certain favorable circumstances, temporarily interdict the terrain. ***They can never occupy it permanently.***

In general, their action, even massive, cannot be sufficient, without the cooperation of the other arms, to break *very solidly organized positions.*

38. Tanks include various types of machines. They are differentiated by their characteristics, namely: speed, mobility, armament, protection, radius of action, and means of command.

Tanks can be assigned the mission of:

— accompanying the infantry in combat and acting in intimate liaison with it by attacking the automatic weapons that stop its advance or;

— largely preceding the infantry and accompanying tank units on their successive objectives, or;

— attacking enemy armored formations, or;

— penetrating deeply into the enemy's disposition ***as soon as it seems sufficiently shaken*** and thus reach the most distant weapons and command units.

When there is no longer any very serious resistance and the exploitation of success seems possible, the tanks can then form the backbone of *mechanized detachments,* constituted in particular by using the armored vehicles of all the units in a position to intervene (motorized reconnaissance battalions, motorized brigades of Cavalry Divisions, and possibly Light Mechanized Divisions).

The slowest tank units and the least suitable for deep actions receive infantry support missions as a priority.

When equipped with radios, the fastest tank units can, in the heat of combat, constitute a flexible and powerful element of maneuver in the hands of a Large Unit commander.

Artillery.

39. Artillery is the arm of *powerful* and *deep fires.*

It constitutes the backbone of the battlefield, due to its stability and lower vulnerability.

For the leader, it is:

— *the instrument of force* par excellence, which allows him to make massive concentrations of fire, thanks to the mobility of its fire plans, the rate-of-fire of its guns and the gamut of its shells and its ranges;

— an available *reserve* because it can, most often, even when engaged, be withdrawn from the front in order to deploy it to other locations.

40. Artillery fire produces psychological and matériel effects ranging from neutralization to the destruction of enemy personnel and their armament.

Artillery performance depends on:

— the organization of a close and constant liaison with the arms for the benefit of which it is employed and with the command;

— its means of signal communications;

— the value of its observation.

Observation, ground and air, has a major influence on the effectiveness of artillery fire.

Lack of observation sometimes imposes significant delays and always leads to very high consumption of ammunition.

The artillery has two primary missions:

— *to aid the infantry and tanks.*

To this end, in the offensive: it prepares and supports the attacks; in the defensive: it delays and breaks up the enemy attacks; it then contributes, with all its fire, to shatter and repel them. In both cases, it conducts counterbattery operations against the enemy artillery and seeks the destruction of armored vehicles;

— *give depth to the offensive or defensive battle* by attacking the adversary's reserves, command units, and communication routes.

For each operation, the command, at all echelons, defines the artillery missions as well as their order of priority, grants an *ammunition allocation,* and generally *distributes its means* into two detachments:

— one placed at the disposal of or working for the benefit of subordinate units;

— the other, the use of which he reserves for himself.

Depending on these decisions, the artillery is articulated in *groupings*.

These groupings are intended to fulfill well-defined missions: direct support and protection of infantry and tanks, counterbattery, interdiction, etc.; however, apart from its regular mission, any group of artillery can be employed, according to the needs, to use its fires to reinforce other groupings, which requires command centralization.

Despite this constraint, the organization of artillery within the framework of any Large Unit must always be flexible. Whenever the circumstances so require: close coordination with the infantry, mobility of the action, extent of the fronts, the organization carried out must allow *decentralization of fires* (general case of direct support, for example) or *decentralization of command* (case of artillery detachments placed under the direct orders of commanders of advance guard detachments or tactical groupings).

41. *Fortress artillery* includes:

— *artillery of the ouvrages,* in casemates or turrets;

— *artillery outside the ouvrages* (*mobile artillery and position artillery)* analogous to the artillery of the large field units.

The artillery of the *ouvrages* is fixed and protected. Its performance is increased by mechanical arrangements. It has armored observation posts, a secure signal communications network, and a supply of ammunition, important but, in the event of investment in the *ourvrage*, limited.

It has, moreover, the same characteristics as the light artillery of the large field units. It produces the same effects. The same rules govern its use.

It is, therefore, incorporated into the mass of artillery operating in the same zone of action, according to the procedures set out in No. 282.

The artillery of the *ouvrages* has an *essential* mission of participating, most often by flanking actions, *in the defense of the main line of resistance.*

The advanced locations and extensive fields of fire of the artillery in turrets allow it, in addition, to execute, to the extent prescribed by the command, *distant action* fires.

42. *Antiaircraft artillery* fights enemy air actions in combination with friendly aviation.

It is armed for this purpose with special cannons, searchlights, and listening dispositions.

Its action is exercised day and night. Thanks to the mobility of its matériel, antiaircraft artillery lends itself to maneuvers and concentrations of means and fires.

Cavalry.

43. Cavalry is the arm of *movement*. It is characterized by its ability to move its means of fire quickly.

The cavalry informs, covers, and fights in liaison with the other arms; it can be called upon to prolong or supplement their action where it is necessary to operate quickly, at a distance, and by surprise.

In collaboration with aviation, it searches for the enemy, determines the situation of its forward elements, maintains contact with them, and provides the command with the intelligence it requests. In general, intelligence can only be clarified through combat.

At the same time, the cavalry covers, either forward of the front, or on the flanks, according to the situation of the Large Unit for the benefit of which it operates.

In combat, cavalry acts through fire combined with movement.

44. The Cavalry includes mounted units, motorized units (motorized dragoons, motorcyclists), and mechanized units (armored cars and tanks).

Mounted and motorized units normally fight on foot.

They are equipped with an armament similar to that of the infantry (light and heavy weapons), which gives them great firepower. They are equipped with tools, as well as means of observation and signal communications and means of crossing rivers. Mounted units are additionally armed with the saber for mounted actions, which are only possible for small units and in special circumstances.

Mechanized units are equipped with armored vehicles armed with machine guns and cannons. They are equipped with radios.

The cavalry forms autonomous Large Units (Cavalry Divisions and Corps, Light Mechanized Divisions), special units (Spahi Brigades), and units that are a part of the composition of the combined arms Large Units (reconnaissance battalions of Infantry Divisions, Army Corps, and Fortified Regions).

The reconnaissance battalions are the security units of their Large Units. During battle, they are most commonly used as a mobile reserve.

The Cavalry Divisions and Corps ensure, for the benefit of Large Units, ground reconnaissance and contribute to their security.

Mobile and extensively equipped with means of fire and armored vehicles, they can intervene in the battle, in liaison with the other Large Units, by rapid and powerful actions or are maintained in strategic reserves (No. 430 and following).

In the offensive, they are the essential instrument for the exploitation of success; in the defensive, they are particularly suitable for a withdrawal from action. Held in reserve, they can intervene to restore the continuity of a front, exploit a success, ward off an enveloping maneuver, or initiate such a maneuver.

Engineers.

45. The mission of engineers is:

— *in the offensive* to create, develop, or re-establish lines of communication according to an order of priority fixed by the command and, in particular, to contribute to the crossing of rivers;

— *in the defensive:*

— to establish or improve the signal communications of the battlefield and its rear;

— to prepare and generally to implement the destruction intended to delay the progress of the enemy;

— to participate in the organization of the site by carrying out certain special works and, where applicable, routine works, either because this work must be carried out within very short deadlines, or because no more important mission has been entrusted to its staff;

— *both offensively and defensively:*

— to ensure the signal communications necessary for the exercise of the command of the Large Units;

— to carry out installation work of all kinds.

Engineers can also be called upon to engage in mine warfare.

46. Engineers collaborate on the battlefield with the other arms, which must generally ensure their protection.

They always work in detachments constituted under the command of its leaders; it directs the units of auxiliary laborers placed at its disposal.

They are equipped with tools, machines, and various matériels for the execution of its technical work as well as the individual armament necessary for the immediate defense of their sites when circumstances require it.

Engineers include a wide variety of units. Some are at the exclusive disposal of the Commander-in-Chief for work of general interest (in particular, pontoon and heavy bridge units); others are temporarily assigned to armies according to their needs at the time (in particular, electromechanics units); still others belong to the Large Units and mainly comprise bridge train companies.

47. The engineering troops (telegraph, radio-telegraph, etc.) that ensure the signal communications of the Large Units are organized in special formations.

The same applies to those (railroad engineers) who ensure the restoration, development, and operation of standard- and narrow-gauge railroads, in collaboration with the field railroad units and the personnel of the networks.

48. In a fortified position, the missions assigned to engineers are generally more complex than in a field position.

The particularities of its use are as follows:

— *outside the ouvrages,* it plays an immediate role, from the start of operations:

— at the front, for deploying obstacles of all kinds prepared in time of peace;

— in the rear for the operation of engineer parks and stores, and the operation of certain railroads.

— *within the ouvrages,* its missions are carried out by specialized detachments:

— sappers who carry out repair work on units damaged by bombardment, repair breaches in obstacles, take part in clearing fields of fire, and possibly carry out counter-mines;

— electro-mechanical engineers who ensure the operation, maintenance, and repair of all electrical and mechanical installations;

— railroad engineers in charge of transport inside the structure.

An officer assistant to the commander of the work commands these detachments.

49. The signal communications of a fortified front are characterized:

— by the wealth of means which not only ensure the necessary relations for the defense of the *ouvrages* but also the links with the reinforcement troops (in particular the artillery) and the Large Units concerned;

— by the security of the installations that facilitate the operation of signal communications.

The latter is served by a special staff of fortress telegraph and wireless engineers.

Air forces and antiaircraft defense.

50. Air forces include aviation and the aerostation.

Aviation scouts, links, covers, and fights.

The aerostation informs.

51. Aviation *intelligence* is characterized by its breadth, its depth across enemy territory, the accuracy of its photographic records, and the speed with which it is obtained or transmitted.

On the other hand, its research is subject to uncertainties and obstacles (covered regions, low visibility of halted and camouflaged elements, atmospheric circumstances, enemy aviation) that do not allow certain conclusions to be drawn from negative intelligence. It is also temporary.

The purpose of the liaisons requested from aviation is to:

— to inform the command about the situation of friendly units;

— to facilitate the cooperation of arms on the battlefield (infantry-artillery);

— transmitting orders or reports between the command echelons.

To ensure these liaisons, aviation has means of signal communications (weighted messages, flares, radios, etc.).

Aviation contributes to the *air cover* of armies by seeking the destruction of enemy air forces either on the ground or in the air and by opposing the crossing of lines by enemy aircraft.

Due to the limitation of its manpower, it is necessary to concentrate its action in the most important areas and at the most necessary times. This action must be combined with that of antiaircraft defense.

52. *Aviation fights*, either against enemy aircraft or ground targets.

Air combat only gradually leads to the final removal of the enemy's aviation.

The possibilities and effectiveness of attacks on ground targets depend essentially on the size of the targets and the antiaircraft reaction of the enemy. Practically, aviation can carry its offensive up to the extreme limit of the radius of action of its apparatuses.

The constant threat it poses over the entire extent of the enemy's territory makes this arm more formidable every day.

The development of motorization opens up new use opportunities for aviation, whether it is to protect the movement of friendly motorized forces or to attack enemy motorized forces beyond the range of our artillery.

Aviation also makes it possible to transport troops and matériel (No. 300).

53. Aviation is organized into detachments of the arm:

Observation aviation informs and joins, either for the benefit of the command, or for the benefit of the other services. The liaison missions inside the lines are entrusted to courier planes.

Light defense aviation covers and conducts aerial combat in combination with antiaircraft defense. In particularly critical moments, it can intervene in the combat on the ground with its machine guns. But this mission exposes it to very rapid wear and should only be ordered to do so in an exceptional circumstance.

Heavy defense aviation attacks battlefield targets as well as enemy communication routes and vital units to the limit of its range. It operates day and night, seeking the only massive actions capable of producing appreciable results.

This specialization of the different aviation categories is neither absolute nor exclusive.

The action of the three subdivisions can, on the other hand, be concentrated on the critical point at a given moment for the success of the operations, particularly in the combat against the enemy air forces.

54. The air forces usually placed at the commanding general's disposal include:

— intelligence units (reconnaissance and observation);

— light defense aviation units.

In addition, at the request of the circumstances, these aviation formations can be reinforced by other formations of the three aviation detachments.

55. The *aerostation* provides, with its balloons and autogyros, aerial observation posts.

The advantages of *balloons* are: extended views, relative stability in space, the possibility of a constant and reciprocal liaison with the ground, and a certain permanence of observation.

On the other hand, the balloon is subjected to atmospheric restraints and constitutes a very vulnerable target to the attacks of the artillery and the enemy aircraft.

56. *The autogyro* has the possibility of landing almost anywhere in the field and of keeping itself, much better than the airplane, in close contact with the command posts.

It can, furthermore, for short periods, reduce its flight to an extremely reduced speed, which allows it to observe in stationary conditions approaching that of the balloon.

On the other hand, in its current state, it cannot receive effective armament, and it remains very vulnerable to enemy aviation but without constituting a target as exposed as the balloon. Its low speed also makes it very vulnerable to automatic weapon fire from air defense.

The autogyro cannot, therefore, be engaged over enemy lines during the day. It is a machine very suitable for carrying out liaisons inside the friendly disposition and for ensuring the observation of the close fires of the artillery by preserving with this arm a more intimate liaison than aviation but less continuous than the balloon.

The autogyro is finally able to fulfill certain night-time reconnaissance missions.

57. *Antiaircraft defense* (D.C.A.)[1] has the mission, in combination with aviation, to oppose incursions by enemy aircraft.

The antiaircraft defense has, for this purpose, artillery units and, possibly, special machine guns, searchlight units, and protection balloon units.

---

[1] D.C.A. = *Défense contre aéronefs* (antiaircraft defense).—Trans.

These units are assigned in varying numbers to the various Large Units according to their needs.

The action of the antiaircraft defense is combined with that of the territorial antiaircraft defense (D.A.T.)[1] stationed in the Combat Zone when the armies operate in national territory. The D.A.T. includes artillery units, special machine gun units, searchlight units, and units of the general security service (watch service, intelligence center, means of signal communications).

Train.

58. The primary purpose of the train is to provide road and rail transport.

It includes horse-drawn, motorized, or mixed units.

These units are equipped with personal weapons and machine guns to help defend the convoys they form against ground and air attacks.

II. Services.

59. The services are responsible for meeting the needs of the command and the troops, ensuring supplies and evacuations of all kinds, as well as the maintenance of personnel, animals, and matériel.

Directors and service chiefs receive orders from the command regarding the use of their service; they ensure its execution and are responsible for it; they can address to the command proposals about this use.

They regulate the operation of their service by complying with the technical instructions given to them by their service at the higher level.

60. Broadly, services fall into three categories:

— supply and maintenance services;

— transportation services;

— maintenance of order services.

These services are organized not only in the Large Units, but also, in whole or in part, in the different zones of the stages.

Supply and maintenance services.

61. *The supply and maintenance* services are the artillery, engineering, signal communications, intendance, health services, aeronautical services, veterinary services, remounts, military postal, and treasury.

62. *The Artillery Service* supplies the troops with ammunition, pyrotechnics, gasoline, and lubricants.

---

[1] D.A.T. = *Défense anti-aérienne du territoire* (territorial antiaircraft defense).—Trans.

It replaces and repairs weapons, vehicles of military crews (as well as harnesses and fittings), motor vehicles (with the exception of special technical vehicles for certain services), and protective equipment against gases.

The Artillery Commander of a Large Unit directs the artillery service in that Large Unit.

63. The *Engineering Service* includes:

— the engineering *matériel service,* which provides troops of all arms with tools, defense matériel, field installation matériel, explosives for destruction, and matériel for crossing rivers;

— five ancillary services, attached to the engineering service and which, moreover, only operate separately at the Army level:

— *Road Service,*

— *Camps and Cantonments Service,*

— *Water Service,*

— *Electrical Service,*

— *Forest Service.*

The Engineering Service is directed in any Large Unit by the Engineer Commander of this Large Unit.

64. The *Signal Communications Service* ensures the supply of troops of all arms and services with telegraph, telephone, radio, carrier pigeon, and signaling equipment (with the exception of pyrotechnics).

65. The *Intendance Service*[1] is mainly responsible for:

— ensuring the supply of food, fodder, fuel, and matériel for the subsistence service, and providing the troops with clothing and encampment kits as well as sleeping matériel;

— authorizing payment and verifying and closing the accounts of the *corps de troupe*[2] within the limits fixed by the regulations.

66. The *Health Service* is responsible for all measures relating to hygiene and disease prevention, treatment of the sick, rehabilitation, triage, hospitalization, and evacuation of the wounded and gassed, all operations that allow the manpower to be maintained or recovered quickly. It ensures the supply of the various formations with health personnel and matériel.

---

[1] Roughly equivalent to the quartermaster service in the U. S. Army.—Trans.

[2] *Corps de troupe:* The largest body in each arm (artillery, cavalry, infantry, etc.) whose command and administration are exercised by one and the same leader, hence, generally a regiment but possibly a battalion.—Trans.

67. The *Aeronautical Service* supplies the aeronautical units with technical matériel for aviation (aircraft, engines, vehicles, etc.) and ballooning; it provides simple repairs.

The supply of aeronautics with armament and special ammunition, gasoline (for airplanes and motor vehicles), and related lubricants is ensured by the artillery service.

68. The *Veterinary Service* ensures the health surveillance of animals, the treatment of sick or injured animals, evacuations, and the supply of medicines and matériel.

69. The *Remount Service* supplies horses, mules, and pack animals. This supply can also come from depots in the interior.

70. The *Armed Forces Postal Service* sends correspondence and postal parcels to the troops and, conversely, forwards these to the interior. It also performs all other authorized postal operations.

71. The *Treasury Service* handles all receipts from the public treasury or made on behalf of the State, and pays the regularly authorized expenses for the *corps de troupe* or the services.

Transportation services.

72. *Transportation services* include bodies whose purpose is the construction, maintenance, and operation of railroads and waterways, as well as the operation of roads. These bodies are:

— the *Railroad Service,* extending to all the rail networks of general interest for the armies (standard gauge and possibly narrow gauge), which the Commander-in-Chief keeps under his immediate control, and whose operation is provided by the Director of Railroads at General Headquarters;

— the *Narrow-gauge Service* (generally 60-cm gauge), organized by an army;

— the *Waterways Service,* under the dependence of the commander of the chief; it is operated by the Director of Waterways at General Headquarters;

— the *Road Transport Service,* ensured, at headquarters level, by the Director of Road Transport, with the train formations (motor and horse-drawn vehicle) that he has kept at his disposal, and on the part of the road network of which he has reserved the operation. The army, with its own means, carries out the transport and operates the roads in its area, taking into account the part that the general headquarters has thus assigned to itself by priority.

Maintenance of order services.

73. The primary *maintenance of order services* are *the Gendarmerie Service* and the *Military Justice Service.*

The *Gendarmerie Service* provides the armies with the general police, the judicial police, and the exercise of provost justice.

The *Military Justice Service* investigates and tries offenses and crimes falling within the jurisdiction of the military courts in the combat zone.

There is also an *Army Security Service* and a *Press Service.*

III. Organization of the lines of communication territory and service.

74. On mobilization, the national territory is divided into an *Interior Zone* (*zone de l'intérieur*) remaining under the authority of the Minister of War, and a *Combat Zone* (*zone d'armées*) placed under the authority of the Commander-in-Chief. The Combat Zone is extended, if necessary, onto the occupied parts of enemy or allied territory.

In the Combat Zone, a distinction is made for each Army between the *Frontline Zone (zone de front)*, assigned to the Army Corps and front-line divisions, and the *Communications Zone (zone des étapes)*, under the orders of the General Director of the Army Lines of Communication. In addition, communication zones may be organized under the direct responsibility of the Army Group Commanders or the Commander-in-Chief. The communications zones cover all the territory between the Forward Zone *(zone de l'avant)* and the rear limit of the Combat Zone.

The Director of the Lines of Communication (*directeur des étapes*) is responsible for exercising command and directing the territorial administration of his zone. His attributions concern: order and security of the territory, police and general discipline, guard of the lines of communication, antiaircraft defense, and exploitation of the local resources. He has a staff, lines of communication commanders, service chiefs and directors, lines of communication troops, and, possibly, other units.

ARTICLE 3.

ORGANIZATION AND GENERAL CHARACTERISTICS OF LARGE UNITS.

THE INFANTRY DIVISION.

75. The Infantry Division is ***the unit of combat,*** within which the action of the different arms is organically combined; it forms a whole that should never be broken up. Its frame allows it to receive a certain number of reinforcing elements.

The Infantry Division is the first unit capable of leading an attack of some importance; however, it has only a restricted field of action and only a limited possibility of lasting.

76. In principle, the Infantry Division consists of the following elements:

— *infantry*: infantry regiments or chasseur[1] demi-brigades, grouped or not in brigades;

— *artillery*: light artillery and heavy howitzers;

— *cavalry*: a Divisional Reconnaissance Battalion;

— *engineers*: sappers-miners;

— *train*: horse-drawn or motorized units;

— *pioneers*;[2]

— *signal communications troops*, observation elements.

It can also receive various reinforcing elements (No. 377).

In certain divisions, notably Motorized Divisions and Mountain Divisions, these constituent elements have a particular organization.

The Infantry Division has the services necessary to meet its immediate needs.

THE CAVALRY DIVISION.

77. The Cavalry Division constitutes a unit provided with means of rapid displacement, horses or motors, and endowed with powerful means of fire. It is characterized by its mobility across all terrains and its speed of maneuver. It can, depending on its mission, receive reinforcements.

The Cavalry Division includes, in principle, motorized or mounted elements, light and heavy artillery, sappers-miners, a bridging train, an air force command and a squadron, courier planes, and signal communications troops.

All of its trains are motorized. It has the necessary services to meet its immediate needs.

Certain divisions receive a special composition: such as the *Light Mechanized Divisions.*

Cavalry Divisions are part of the Commander-in-Chief's reserves; they can be grouped into Cavalry Corps or enter into the composition of an Army, an Army Detachment, or an Army Corps.

---

[1] Light infantry trained and equipped for rapid movements, includes *chasseurs à pied* and Alpine mountain troops (*chasseurs Alpins*)—Trans.

[2] Also called sappers-pioneers, these are troops used to build field fortifications, roads, bridges, etc. and to create breaches in enemy defensive positions. They are part of the infantry and not the engineers.—Trans.

THE ARMY CORPS.

78. The Army Corps is ***the unit of battle,*** capable of carrying out, within the framework of the Army, extended tactical action and leading it to the decision.

Its commander has organic elements of the Army Corps (reinforced or not by elements taken from the General Reserves) and a variable number of divisions. He leads and coordinates the combat of his divisions and intervenes directly in the fight by extending or reinforcing the action of some of them using elements kept in reserve or made available. He directs the use of his artillery.

Equipped with solidly constituted command bodies, the Army Corps is capable of supervising a large number of reinforcements, isolated divisions, or troops of the various arms and ensuring their implementation in battle; it can thus adapt to the varied missions entrusted to it.

79. The main task of the *Army Corps Artillery Commander* is to direct and coordinate, in accordance with the instructions of the Army Corps commander, and the fires of the whole of the Army Corps artillery.

This artillery includes:

— the non-divisional artillery composed, organically, of the heavy artillery of the Army Corps, and, possibly, reinforcement units;

— divisional artillery and its reinforcement units.

The distribution of the reinforcement units between the non-divisional artillery and the divisional artillery is carried out by the Army Corps artillery commander, in accordance with the decision of the commander.

The Army Corps artillery commander also ensures the command of the non-divisional artillery.

Finally, he directs the supply of artillery and infantry ammunition for the entire corps and is responsible for it.

80. The *Commander of Engineers of the Army Corps* exercises command of the engineer units, organic or not, placed under the direct control of the Army Corps.

He ensures, in accordance with the instructions of the Army Corps Commander, the distribution of reinforcement engineering units.

He coordinates, according to the technical directives of the Commander of Engineers of the Army, the action of the engineers of the divisions and the Army Corps whenever necessary (in particular in the lines of communication construction and maintenance).

Finally, he ensures the supply of engineering matériel and tools to all units of all arms and services attached to the Army Corps.

81. *The Commander of the Air Forces of the Army Corps* commands the aeronautical units, organic or of reinforcement, placed at the disposal of the Army Corps; he satisfies the demands of the divisions and the heavy artillery according to his availability and according to the order of priority set by the Army Corps Commander.

82. The *Commander of Tanks of the Army Corps*, designated when this Large Unit is provided with tanks, ensures, in accordance with the orders of the Army Corps Commander, the distribution of tank units between the divisions.

He commands the tanks that the corps commander keeps at his disposal.

83. The corps includes typically, in addition to divisions, the following organic elements, whether motorized or not:

— *artillery:* long-range heavy artillery;

— *cavalry:* an Army Corps Reconnaissance Battalion;

— *engineering:* sappers-miners;

— *air forces:* observation, liaison and the aerostation;

— *train:* one motorized unit and one horse-drawn unit.

The corps includes, in addition, pioneers, signal communications units, observation units, and bridging equipment.

The organic elements can be temporarily reinforced in artillery, engineers, air forces and antiaircraft defense, tanks, and means of transport.

As a general rule, the services of the Army Corps, having above all a role of direction and coordination vis-à-vis the corresponding divisional services, mainly include administration bodies. The implementation bodies of the services of the Army Corps satisfy, in principle, only the particular needs of the organic elements of the Army Corps.

THE CAVALRY CORPS.

84. The Cavalry Corps comprises a variable number of Cavalry Divisions, to which other combat elements may be added: infantry, tanks, artillery, engineers, air forces, services, etc.

The Cavalry Corps has properties similar to those of the Cavalry Divisions, but with larger command bodies, superior firepower, and more developed maneuver and combat possibilities.

The formation of the Cavalry Corps is only contingent; it makes it possible to place under the same authority, when the need arises, units possessing similar properties, employed on the same ground for the same mission.

THE ARMY AND ARMY DETACHMENT.

85. The Army is the ***fundamental unit of strategic maneuver.***

Its leader combines and directs the maneuver of several Army Corps, Large Cavalry Units, and air forces within the framework of the same overall mission.

The Army is a command and control body. It has, in principle, as organic elements, only a headquarters comprising a staff, commands of certain arms and directors of services, specialized troops, etc.; but it receives organic elements of Army Corps and Cavalry Corps, Infantry Divisions, Cavalry Divisions, Light Mechanized Divisions, Fortified Region Mixed Brigades, units of artillery, engineers, air forces, tanks, and service elements in variable numbers depending on its mission, etc.

The army reports to the Commander-in-Chief, either directly or through an Army Group Commander.

86. The *Chief of Artillery of an Army* ensures, in accordance with the instructions of the Army Commander, the distribution between Army Corps of the artillery of the General Reserve assigned to an Army and exercises the command of the artillery maintained at the direct disposal of the Army. He coordinates the action of the corps artillery among themselves and with the army artillery. Finally, he exercises the senior administration of the supply of ammunition, gasoline, and various matériels for the artillery service.

87. The *Chief of Engineers of an Army* exercises command of the engineer units, organic or not, kept at the direct disposal of the Army Commander.

He ensures, in accordance with the instructions of the Army Commander, the distribution of the reinforcement engineering units.

He coordinates the actions of the engineers of the divisions, the army corps, and the army whenever necessary (particularly in building and maintaining communication routes).

Finally, he manages the engineering service and, in particular, the supply of matériel.

88. The *Chief of the Air Forces and Antiaircraft Defense of an Army* exercises the command of the air and air defense formations, organic or not, maintained at the direct disposal of the Army Commander.

In this respect, the Army Commander has, in his zone (frontline zone and communications zone):

— all antiaircraft defense formations assigned to the army (No. 96);

— all territorial antiaircraft defense formations stationed in the Combat Zone, provided that the territorial antiaircraft defense posture is not altered without prior authorization from the commanding general; these formations are made available to the generals commanding the armies as soon as they take command.

The Commanding General of an Army is, moreover, in charge of the administration of all passive defense measures in his Combat Zone (No. 304).

The Commander of the Air forces and Antiaircraft Defense of the Army ensures, in accordance with the instructions of the army commander:

— the distribution of reinforcement air units;

— the combination of the various means of air defense (territorial antiaircraft defense, defensive light aviation, and antiaircraft artillery).

He manages the supply of the air formations of the army and the large units that compose it.

89. The *Chief of Tanks of an Army* ensures, in accordance with the instructions of the Army Commander, the distribution of tank units between the Army Corps and exercises command of those held in reserve by the Army Commander. He regulates the special supply to the tanks as well as the evacuations and the repairs.

90. The Army has many powerful *service units.* The administration bodies are strongly represented there. The implementation bodies include significant means likely to be reinforced when the situation of the army requires it.

The main implementation bodies consist of:

— depots and establishments (miscellaneous supplies, parks, stores, workshops, hospitals, etc.), belonging to the various services;

— motorized or horse-drawn means of transport, possibly reinforced by narrow-gauge railroads;

— technical units belonging directly to the various services and by laborer units.

91. *The Army Detachment* is a temporary grouping, created to deal with a special situation.

Its manpower varies according to its mission. It has a headquarters and, possibly, troops and organic services corresponding to the large units entering into its composition.

THE ARMY GROUP.

92. When several armies are operating in the same theater of operations, one or more *Army Groups* (*groupes d'armées*) may be formed, each of them comprising the armies assigned to the same compartment of the theater of operations or contributing to the same mission.

The Army Group Commander directs and coordinates the operations of his armies (possibly Large Cavalry Units and air force units assigned to him), in accordance with the operations plan drawn up by him, within the framework of his mission.

93. The Army Group is exclusively an **organ of command;** its mission is, above all, of a strategic nature. Its leader has a reduced headquarters.

The Army Group Commander may receive, in addition to the armies of his group, Large Units or units from the General Reserves that he keeps, as the case may be, at his disposal or which he assigns to his armies. The Large Units kept at the disposal of the Army Group Commander are attached for administration and maintenance to one of the armies of the group.

In addition, the Army Group has an Army Group Air Force Command, responsible for the coordination of the army air forces and, possibly, the implementation of the air formations that would be assigned to the Army Group.

In principle, the Army Group does not have services.

In some cases, he may be assigned a communication zone directorate (*direction des étapes*) to command and administer a part of the Army Group's communication zone.

THE COMMANDER-IN-CHIEF.

94. All the forces operating in the same theater of operations or in several closely related theaters are placed under the command of a Commander-in-Chief.

Every Commander-in-Chief has a staff, inspection bodies, and service administration bodies. The inspectors are the technical advisers of the Commander-in-Chief; they receive from him missions of inspection, investigation, and control. The directors allow him to ensure the administration and maintenance of his forces and the exploitation of the railroads, waterways, and roads of his theater of operations.

The whole of the staff, the inspectorates, and the services at the disposal of the Commander-in-Chief for the exercise of his command constitute the General Headquarters.

In addition, the Commander-in-Chief has, for certain arms, General Reserve command bodies (tanks, artillery, train, engineers).

95. The Commander-in-Chief in a theater of operations has all the forces assembled in that theater at his disposal for the execution of his plans of operations.

The Commander-in-Chief thus has the possibility of equipping exactly, using the General Reserves, the Army Groups, or the Armies according to the mission which falls to them, and in particular to practice on the passive fronts a severe economy of forces to the advantage of fronts on which he intends to seek a decision. He may retain, to be employed under his direct orders, specific aviation units.

96. In the Combat Zone, the entire air defense, with the exception of that of the seaports, is under the orders of the Commander-in-Chief.

The Commander-in-Chief has all powers in the triple domain of active defense, passive defense, and general security.

He has the air resources assigned to the armed forces and those made available to him by the government.

He cannot, however, modify the general system of the territorial antiaircraft defense and, more specifically, that of the general security service, without agreement with the Commander-in-Chief of the Air Forces.

The authority of the Commander-in-Chief is exercised through the Army Group and Army Commanders.

The combination between the air defense of the armies and the territorial antiaircraft defense is ensured:

— in the zone of each Army, by the Commander of the Air Forces and the Antiaircraft Defense of the Army, in accordance with the instructions of the Army Commander;

— in the rest of the Combat Zone, by the Commander of the Air Forces and the Antiaircraft Defense of the Armies, in accordance with the instructions of the Commander-in-Chief.

97. When the theater of operations extends outside the national territory, special measures are prescribed to ensure the command and administration of the occupied countries and to determine relations with the local authorities.

The purpose of these measures is to ensure, in the first place, the freedom of action necessary for the Commander-in-Chief in this theater of operations, the security of his troops, and the proper functioning of his communication routes.

ARTICLE 4.

FORTIFIED REGIONS. — FORTIFIED SECTORS AND DEFENSIVE SECTORS.

98. Fortified organizations increase the power of armies. ***They are integrated into their general system.***

99. A *Fortified Region* is an extended defensive complex which includes:

— a position of resistance reinforced by fortifications and equipped with powerful weapons;

— equipment of the front with numerous and well-protected means of command;

— equipment of the rear areas prepared in peacetime;

— supplies of all kinds.

100. Because of their initial mission, the Fortified Regions are organized in sensitive border zones that are particularly favorable to the penetration of a powerful and rapid enemy offensive; in these zones, they are as far as possible drawn in such a way as to cover the industrial centers and the parts of the territory to be kept for the development of operations.

101. The Fortified Regions, contrary to the old fortresses, ***do not constitute closed systems.*** To support the battle, a fortified unit, like the other elements of the armies, will have to make a constant call on the resources of the national territory.

The front of the Fortified Regions is moreover always very extensive, so that, even overwhelmed on their flanks, these regions can maintain continuous relations with the country's interior for as long as possible.

The Fortified Region possesses considerable defensive power, which characterizes its role as a strategic instrument (No. 15).

102. A Fortified Region comprises several *Fortified* or *Defensive Sectors.* Autonomous Fortified Sectors are also created, either in the intervals of Fortified Regions, in border zones covered by a major obstacle, or in compartmentalized zones, particularly in mountainous terrain.

The *Defensive Sectors,* possibly created in the intervals of the Fortified Regions or Sectors, include an organization approaching that of the Fortified Sectors, although less complete.

103. Fortified Regions, Fortified Sectors, and Defensive Sectors are occupied by troops generally organized in Mixed Brigades, which have a composition similar to that of an Infantry Division. Specific defensive sectors can receive a unique organization.

The Fortified Region has organic elements similar to those of an Army Corps.

As soon as all or part of the front of the Fortified Regions or Sectors is threatened by a powerfully mounted attack, it may be necessary to engage field troops on this front.

104. The Fortified Regions, the Fortified Sectors, and the Defensive Sectors, with their organic elements (troops and services), ***are placed under the authority of Army Commanders.***

However, except by order of the Commander-in-Chief, the troops cannot be moved away or distracted from their special mission of defending the organizations entrusted to them. No deductions are made from their matériel or their supplies except in an emergency, and subject to the *ouvrages* are left with ample security supplies. The Army Commanders report to the Commander-in-Chief on the levies ordered by them.

105. In the Fortified Regions and certain Autonomous Fortified Sectors, the various supply and maintenance services have missions identical to those they must fulfill vis-à-vis the other Large Units. Their operation is adapted to the essential characteristics of the training courses served, namely:

— special armament;

— relative fixity of the installations carried out;

— a comprehensive organization of these installations;

— a developed communication network;

— a certain independence of supplies in relation to shipments from the rear because of the importance that can be given to the storage facilities set up in the shelter of the fortified front.

ARTICLE 5.

GENERAL RESERVES.

106. The General Reserves available to the Commander-in-Chief include:

— possibly the staffs of an Army Group, Army, or Cavalry Corps Commander temporarily placed in command reserve;

— Army Corps and Infantry and Cavalry Divisions;

— units of various arms which organically constitute the *General Reserves of Arms;*

— elements of certain services that constitute the *General Reserves of Services.*

CHAPTER II.

MODES OF ACTION.

107. The two modes of action are offensive and defensive. The maneuver usually involves a combination of offensive and defensive forms.

ARTICLE 1.

THE OFFENSIVE.

108. The offensive is the mode of action par excellence.

It consists of forward movement, undertaken with the most potent means possible, to reach the adversary and inflict losses heavy enough to ruin his morale and matériel strength to force him to yield the ground and render him harmless.

Only the offensive makes it possible to obtain these decisive results.

109. The offensive supposes an *initial superiority* that can be made up of various elements: manpower, armament and matériel means, moral, valor or training, strategic situation, priority of preparation, etc.

It requires the implementation of all the resources of the art and science of the leader; it also requires quality troops.

110. The offensive takes multiple forms.

If the front of the adversary leaves open spaces, it calls for maneuvering. If it is continuous, the power factor becomes preponderant.

In both cases, but to different degrees, the element of surprise plays an important role.

Faced with a continuous front, the desired effect of surprise will aim, above all, to keep the adversary uncertain of the moment when the attack will be started, its general direction, and the scale and the unforeseen nature of the means implemented.

In open terrain, in contrast, surprise is a major factor in the success of the maneuver.

ARTICLE 2.

THE DEFENSIVE.

111. Generalized or localized defense is the attitude temporarily chosen by a leader who does not consider himself to be in a position to take the offensive in the entire or certain parts of his zone of action.

He then uses the combination of fire and terrain, generally reinforced by a permanent or improvised organization, to block the enemy's route.

This attitude cannot bring decisive results; as soon as the inferiority that motivated it ceases, it is in the offensive that the commander must seek to win the trial of the enemy forces.

112. Passive defense, consisting solely of resisting enemy attacks and breaking them, has a demoralizing effect on the combatant in the long run; it increases the boldness and confidence of the enemy.

If the defensive must be extended, it will be accompanied at least by offensive actions endowed with the maximum power and means, the union of which will be made possible by the economies of forces practiced on the quiet parts of the front.

These actions will make the adversary cautious and encourage him to maintain manpower in front of the defensive front.

113. Defensive action, which generally takes the form of defending a single position, can be organized and moved successively to several positions; it can significantly delay the enemy's advance and gain time, but with a loss of ground.

Sometimes, the occupation of each successive position will cease before the enemy has completed his preparations for attack; other times, the forces occupying the position will offer initial resistance, then withdraw, avoiding full engagement.

However, these operations are tricky to execute, especially in the presence of armored vehicles. They correspond to specific situations or needs, and the rule remains to defend a chosen position to the full.

CHAPTER III.

ELEMENTS OF ACTION.

114. The action of the armed forces is manifested by fire and movement.

ARTICLE 1.

FIRE.

115. Fire is the dominant factor in combat. It destroys or neutralizes the enemy. The attack is fire that advances; the defense is fire that stops.

Artillery, infantry and tanks, cavalry and armored vehicles, and aviation use the weapons and projectiles at their disposal throughout the battlefield, *making the most of their range and direction of fire*

The command, at all levels, has a constant duty:

— to achieve an ***economy of fire*** in the same way and the same sense as the economy of forces by devoting the greatest amount of ammunition available to the most necessary and most effective fires;

— to seek the best output by the close combination and, in some instances, by superimposing fires from all sources, coordinated in a single system organized and regulated by the commander (fire plan).

The effects of fire are both matériel and psychological. They create zones of death where troops suffer massive and staggering losses that render them helpless and pin them to the ground, where matériel is destroyed, and where organizations are disrupted.

Fire systems are not generally the same day and night.

116. *In the defensive,* a continuous and deep system of fires, prepared, installed, and combined with the organization of the terrain, opposes the adversary with a formidable stopping power.

Such a front can only be broken through by using numerous powerful means that require a long time to bring together.

117. *In the offensive,* the crushing and neutralizing effects of preparation fire destroy obstacles and benefit the debouchment.

The combined fires of the different arms accompany and protect the attacking troops. The observation of fire and the functioning of a safe liaison between infantry and artillery are the conditions of their effectiveness.

In any case, the power of fire depends on the prior assembly and continuous renewal of ammunition supplies.

118. To the effects of fire can be added the action of smoke, mines, and possibly poison gas, if their use has been ordered (see Foreword).

The use of smoke makes it possible to blind enemy observation posts and to cover the placement, departure, and advance of attacking troops.

Mines are used on stabilized fronts. They serve to disrupt the enemy's advanced organizations and counterattack their underground approaches. Their psychological impact is undeniable; however, their matériel effect is rarely proportionate to the required efforts and sacrifices.

Operations that are based on the use of these mining operations are generally called "mine warfare."

In mobile warfare, as on stabilized fronts, special weapons, known as *antitank mines*, are used to deny certain areas of terrain to armored vehicles.

Chemical agents (*gaz de combat*) provide the means either to put the personnel out of action or to reduce the resistance of the personnel by forcing them to wear special protective items. They can deny access to large areas by infecting the ground and objects in a long-lasting way. Their action of neutralization is considerable, on the artillery in particular.

The gases are most often implemented in the form of toxic projectiles launched by artillery or aircraft, more rarely by emissions of gas clouds intended to invade the enemy positions.

The useful effect of gases greatly depends on atmospheric conditions.

ARTICLE 2.

MOVEMENT.

119. Forward movement brings the means of fire capable of breaking the enemy's resistance ever closer.

It is the sign of the superiority acquired by the one who realizes it; it enhances the morale of the combatant who advances. Conversely, the one forced to fall back is gradually penetrated by the feeling of failure or the uselessness of his efforts and, finally, of his inferiority.

Movement is one of the essential elements of any maneuver.

Its effects depend on the magnitude of the forces that take part in it and the speed with which it can be carried out.

Speed, being itself the surest guarantor of maintaining secrecy, promotes surprise and, through it, success.

120. The development of motorization and all-terrain vehicles, as well as signal communications methods, opens up maneuvering possibilities

that the size of the masses to be used had caused it to lose. It makes it possible to envisage a certain acceleration of the rhythm of the battle.
It becomes possible:

— thanks to motorization, to transport considerable forces quickly and far;

— through the use of mechanized units or detachments, to rapidly provide screening, if necessary, to motorized transport;

— thanks to armored vehicles, to accelerate movement in all phases of offensive or defensive combat.

Motorization and mechanization also facilitate the resupply of large and small units.

TITLE III.

INTELLIGENCE AND SECURITY.

CHAPTER ONE.

INTELLIGENCE.

ARTICLE 1.

GENERAL.

121. Intelligence is necessary for any Large Unit commander to enable him to:

— on the one hand, to conduct his own *maneuver,* taking into account the enemy's situation at all times;

— on the other hand, to thwart the adversary's initiatives in time and to counter them to guarantee his *security.*

It is important for him to obtain intelligence on:

— the presence or absence of the enemy at a given time in a given area;

— his general attitude (forward movement, immobility, or retreat);

— his apparent outline and the identification of its first-echelon elements; the situation, movements, and locations of its main bodies and his reserves;

— his primary zone of action or his primary zone of resistance.

122. The search for intelligence is organized according to the general situation, the intentions of the commander, the possibilities of the intelligence body implemented and taking into account:

— the time required to transmit the intelligence;

— the expected speed of movement of the adversary's means of combat and, consequently, the size of the movements that he will be able to carry out;

— the time needed to modify the system based on the intelligence collected.

These various considerations lead, in the current state of motorization and signal communications methods, to seek intelligence at great distances, which are all the more considerable as the higher the level of command concerned.

123. It is up to the command of a Large Unit to ensure the *coordination* of intelligence-gathering efforts and their *constant adaptation* to the diversity of successive situations.

For this purpose, for each operation, often for each phase of the same operation, he makes known its intelligence needs depending on the situation.

These data serve as the basis for drawing up the *intelligence plan* (No. 23), which contains the list of intelligence requested by the commander, either to support his decision or maneuver, or to guarantee his security.

To any intelligence plan, identifying the goal to be achieved and the conditions of time and place in which the intelligence must be provided, corresponds to a *research plan*, the purpose of which is to distribute the research work between the different bodies likely to provide intelligence.

The implementation of this last plan is ensured by using *execution orders* addressed to each of these bodies.

ARTICLE 2.

INTELLIGENCE BODIES. — THEIR POSSIBILITIES.

124. Intelligence is requested from the air forces (aviation and aerostation), cavalry, troops in contact, and special bodies.

125. *Aviation* plays a leading role in intelligence gathering. By the speed, the ceiling, and the radius of action of its dispositions, by its endowments in radios which ensure the immediate transmission of its messages, it constitutes par excellence the long-range reconnaissance arm. Photographic documents lend authenticity and precision to intelligence.

However, aviation does not have a sufficient number of aircraft to keep in the air permanently; furthermore, rain and fog at night make flights difficult and visibility precarious; the presence of the troops in the woods almost completely escapes the airplanes; finally, the air force is generally unable to determine the apparent contour of the front held by the enemy.

The *aerostation*, as soon as the armies come into contact, completes the intelligence provided by the air force.

Its observation is continuous but is limited to the foreground of the battlefield.

The *cavalry* provides continuous observation and operates at all times and in all terrains, especially in close terrain where the troops escape aerial investigations; it takes prisoners. The negative intelligence it provides offers every guarantee of accuracy. It makes it possible to determine the apparent contour of the front held by the enemy and to maintain contact. Finally, it will likely occupy the terrain at least for a certain time.

126. During battle, the dogged search for contact and the combat itself provides the most valuable and reliable intelligence. All troops have a *compelling and continuing duty* to seek intelligence on enemy strength, locations, and movements and to forward it to higher command by the quickest means.

The main sources of this intelligence are: direct visual observation, the interrogation of inhabitants and prisoners, and captured documents.

In order to obtain information, commanders may prescribe limited operations on specific fronts. These are most often carried out in the form of *coups de main.* (No. 406).

127. Special bodies include specialized ground observation bodies, listening bodies, secret services, etc.

ARTICLE 3.

RECONNAISSANCE.

I. Aviation and Large Cavalry Units.

128. The purpose of *reconnaissance*[1] is to provide higher command with the intelligence it deems necessary to develop its plan of maneuver.

It is ensured concurrently by aviation and the Large Cavalry Units.

Aviation seeks the most distant intelligence, which makes it possible to appreciate the situation as a whole. Its effort is particularly focused on the movements of the adversary's main bodies and reserves.

When given sufficient space, cavalry, through *ground reconnaissance,* supplements air reconnaissance by providing closer but accurate intelligence, verified by contact and often tested by combat.

---

[1] The French Army terminology of this period used three different words for what is called reconnaissance today. The word "*reconnaissance*" was used strictly for short-range tactical reconnaissance, what the U.S. Army refers to as "close reconnaissance." Long-range strategic and operational reconnaissance, which the U.S. Army refers to as "distant reconnaissance," was called "*découverte*" ("discovery"). The entirety of the reconnaissance efforts, both close and distant, was called "*exploration.*" In this translation, I have adopted the U.S. Army's terminology in general, but note the Reconnaissance Battalions mentioned in the text were officially designated "*groupes de reconnaissance,*" i.e., "close" reconnaissance battalions.—Trans.

129. *Air reconnaissance* is carried out by the air forces of the Large Unit or of the Large Units of the higher echelon. It encompasses the entire depth of the theater of operations or such zone determined by the command.

A specific order, valid for a period of time that shrinks the faster the situation evolves and the lower the level of command, objectively indicates to the air force the intelligence requested by the command, specifies its nature and order of priority.

Aviation proceeds, in general, by sending reconnaissances and taking photographs.

130. *Ground reconnaissance* is entrusted to the Large Cavalry Units, which receive, for this purpose, indications from the command on the conditions of execution of their intelligence mission in time and space (No. 441).

Large cavalry units conducting *reconnaissance* search for information through a set of elements that constitute the *distant reconnaissance.*

This distant reconnaissance includes, operating in intimate liaison, a *distant air reconnaissance* ensured by the organic aviation of the Large Cavalry Unit and a *distant ground reconnaissance* carried out by reconnaissance and detached light horse-drawn or motorized parties at great distances on the main roads and on a broad front.

The main body of the Large Cavalry Unit advances oriented and informed by distant reconnaissance and able, following the orders of higher command, either to seek contact with large enemies through combat or to guarantee the security of the Large Units with which it cooperates.

II. Reconnaissance Battalions.

131. The main mission of reconnaissance battalions, which are Army Corps or divisional units, is to provide information to the commanders of Large Units at the distances and in the time they deem necessary for their operations.

If their Large Unit is preceded by Large Cavalry Units, the reconnaissance battalions liaise with the latter, ready to take over the task of gathering intelligence should the Large Cavalry Units uncover their Large Unit's front.

If the latter is not preceded by Large Cavalry Units, the reconnaissance battalions alone ensure, within the limits of their means, the search for intelligence.

CHAPTER II.

SECURITY.

ARTICLE 1.

GENERAL.

132. *Security* responds to a permanent need: its purpose is to enable the commander to make arrangements and to ensure the protection of the troops against surprises.

Security is essentially based on intelligence and the *disposition.*

133. The development of the power and range of weapons, the progress of mechanization which favors the deep incursions of armored vehicles and those of motorization which allow the rapid movement of large forces, the increase in speed, and the power and range of aviation today give the notion of security a new and broader value and lead to a distinction between *ground security* and *air security.*

ARTICLE 2.

GROUND SECURITY.

134. The purpose of ground security is to protect Large Units against enemy ground actions.

For this purpose:

— the ***command*** must be able at all times to have a sufficient security zone to maintain its *freedom of action,* i.e., to be able to prescribe in time to the Large Unit the offensive or defensive maneuvers corresponding to the situation.

This is the purpose of ***distant security***;

— each ***Large Unit*** must also have the time and space necessary to allow its main body to make its combat arrangements and to ensure the implementation of its means of fire.

This is the purpose of ***close security;***

— finally, all troops must, at all times and in all places, protect themselves against ground surprises and poison gas.

This is the purpose of ***local security.***

Close security and *local security* constitute the *security of the troops,* as opposed to *distant security,* also called the *security of the leader* (*sûreté du chef*).

135. The ground security of motorized Large Units is subject to specific procedures set out in Title X (No. 455).

136. Ground security is based:

— on *intelligence* (see Chapter One of this Title);

— on the *disposition,* which allows, by a suitable echelonment in depth of the various elements, to quickly counter enemy attempts with a front behind which the commander will have the time to make his arrangements.

The Large Units, on the march or on station, generally present the following echeloned system:

— distant security elements;

— close security elements;

— elements in charge of the maneuver (main body of forces);

— *on a judicious use of the terrain,* thanks to which the troops can avoid the observation and the fires of the adversary (hidden positions, cover) and protect themselves against the incursions of armored vehicles.

The latter, by the matériel and psychological effect produced by their intervention, constitute a formidable factor of surprise against which the command and the troops must always be protected.

Security is, therefore, only guaranteed insofar as the attack or defense dispositions are protected from the irruption of armored vehicles.

This consideration exerts an essential influence on the choice of the positions to be occupied by all security elements.

Rivers, canals, and ponds constitute the most effective obstacle against armored vehicles because of the facilities they offer for defense, the inexpensive destruction they allow, and the difficulty of crossing them when the parts accessible to amphibious vehicles are closely monitored.

Failing these, the choice will be natural obstacles (wooded areas, embankments, cliffs, marshy bottoms) or artificial ones (embanked or sunken roads or tracks, villages, or built-up areas).

I. Distant Security.

137. The distant security interest is guaranteed when:

— the leader is *informed* about the situation and the movements of enemy forces, at a distance commensurate with the size of his Large Unit and the adversary's movement capacity. (No. 122).

— the Large Unit is covered against fast-moving enemy units.

138. Distant security is the combined work;

— aviation;

— the Large Cavalry Units or Reconnaissance Battalions or even combined arms detachments, generally motorized or mechanized, formed when needed.

Depending on the circumstances, these Large Units, units, or formations carry out distant security missions alone, in combination with each other, or in reinforcement of each other.

139. Aviation seeks *intelligence* through air reconnaissance (No. 129). The cavalry (or the detachments that replace or reinforce it) seek *intelligence* through ground reconnaissance (No. 130) and also provide *screening* for the Large Units for whose benefit it operates.

To this end, the cavalry, to fulfill these two missions, deploys two sets of detachments in front of its advance (or flank) guards:

— distant reconnaissance detachments, responsible for the search for intelligence (No. 130);

— distant security detachments, of variable numbers pushed to distances that essentially depend on the terrain, the situation, and the size of the troops to be covered and responsible for ensuring a first defense of the easiest lines of terrain to hold, especially against armored vehicles.

If the situation requires it, the advance guards and then the main body of the cavalry units take responsibility for the defense of the line held by the distant security detachments.

II. Close security.

140. Close security is guaranteed when, on station, on the march, or in combat, the main body of the Large Units is:

— *protected* on its front, on its flanks, and possibly on its rear, against surprises, in particular against the effects of infantry and light artillery fire and against attacks by armored vehicles;

— *informed* about the possibility of such actions.

141. *Close security* is ensured, with the assistance of Reconnaissance Battalions, by combined arms detachments (advance guards, outposts, rear guards, flank guards).

Far from the enemy, the Reconnaissance Battalions communicate to the advance guards (outposts, flank guards) useful intelligence for their maneuver.

Near the enemy, and at the time of establishing contact, they operate in close liaison and for the benefit of these same detachments.

142. *Security in station.* — The main body of the troops is stationed as far as possible away from major obstacles, particularly large terrain breaks, in order to protect themselves against the incursions of the enemy's armored vehicles.

The protection of the main bodies is guaranteed by detachments that take the name of *outposts.*

Away from the enemy, the outposts guard the access routes usable by the enemy. They hold the important points, in particular the crossroads and the obligatory passages of the terrain breaks.

Close to the enemy or in contact with him, the leader, after defining the position of resistance on which, in the event of an attack, he will accept combat, arranges the outposts in such a way as to protect the troops occupying them from an attack by surprise and organizes the position of the resistance.

The mission of the troops at the outposts varies according to the circumstances (No. 252).

143. *Security on the march.* — Security on the march is based on the use of combined arms detachments under the command of the main body, both in terms of time and distance. Depending on the place they occupy in relation to the main body, they are referred to as *advance guards, rear guards,* and *flank guards.*

144. *Advance guards.* — Advance guards are *reconnaissance* and *protection* units that the Large Units detach in front of them to guarantee the close security of their main body.

The command sets their number and their strength in order to put them in a position to reconnoiter in detail the zone of action of the Large Unit in all its width and, when the time comes, to constitute a defensive front under the shelter of which the main body can prepare for battle.

Advance guards include: strictly necessary infantry and engineer units, reinforced, as far as possible, by antitank elements, armored vehicles, and possibly artillery fractions. If the situation requires it, they must be able to be supported by all the artillery of the Large Unit, deployed, however, in such a way that its cover is assured.

The progress of the advance guards is carried out initially on axes, by successive bounds, and under the cover of the distant security units; it is always subordinated to the advance of the main body.

In front of an enemy in position, the advance guards, after rejoining the cavalry that has stopped in contact, push their advance, with the support of increasingly more numerous artillery, until they meet a continuous resistance, impossible to overcome by their own means.

In front of an enemy on the offensive march, the advance guards advance while being ready to deploy quickly with the support of the divisional artillery on the position that the leader has chosen to stop the adversary before attacking him.

In both cases, if the enemy exerts vigorous pressure on them, they attach themselves to the terrain and, reinforced if necessary, oppose the enemy with a solid front that enables the commander to make his arrangements according to his plan.

The advance guards also have the *permanent mission* of contributing to the antitank defense of the Large Unit. To this end, antitank weapons are distributed throughout the depth of their columns and the bounds chosen from among the lines most favorable to this defense.

As a general rule, the commander of the Large Unit personally directs the action of the advance guards.

145. *Rear guards.* — In the retrograde march, undertaken deliberately or under pressure from the enemy, the rear guards have the task of informing and covering the main body in order to enable it to avoid combat.

The composition of the rear guards depends on the nature of the terrain and the time they have to maneuver. In general, they consist of combined arms detachments advantageously composed of motorized or mechanized elements and equipped with antitank weapons.

They operate on axes, in fairly wide bounds, from obstacle to obstacle: natural breaks, rivers, failing which there are obligatory crossing points.

They protect themselves on each of their lines by destructions beaten by artillery for as long as possible and slip away, preferably during the night, leaving only fast elements in contact.

Forced to maneuver during the day, they try to hide their withdrawal by using extensive cover.

By night or by day, aviation impedes, by powerful actions, the enemy's progress.

The security of his flanks is always one of the main concerns of a rear guard commander.

146. *Flank guards.* — The flank guards have the task of providing intelligence and covering the main body of a troop on the march or stationed on its flanks.

Their role acquires special significance due to the extensive operational range of armored vehicles, which heightens flank insecurity.

The flank guards are made up of combined arms elements, preferably motorized, equipped with antitank weapons, and scouted, if possible, by aviation.

Depending on the situation or the nature of the terrain, the flank guards can be fixed or mobile. The former operate defensively, preferably holding natural obstacles; the seconds also can only cover effectively, in the event of an encounter with the enemy, by opposing him with a defensive front.

147. *Combat Security.* — In combat, security is mainly based on the disposition of forces, on maintaining contact, and on the liaison between the different elements.

148. *In the offensive,* security in combat results, first of all, from the continuity of the fires of the first echelon as well as from the powerful fire actions that prepare and accompany the attacks.

In addition, the echeloned combat disposition ensures, to a certain extent, cover for the artillery, the reserves, and the rear.

The protection of exposed flanks is the responsibility of units receiving an appropriate mission.

In the defensive, security in combat results from the fact that the troops, covered by a system of outposts and installed in a position of resistance, are at all times in a condition to face an attack.

In both cases, the proper functioning of the liaison between arms and between the different units of the arms effectively contributes to security.

149. In combat, *maintaining contact* is one of the most important factors of security; it constitutes a permanent obligation for front-line units.

It is up to the leader of these units to immediately notify the higher echelon as soon as the search for or resumption of contact exceeds their possibilities.

First-echelon Large Unit commanders are directly responsible for maintaining contact.

III. Local security.

150. The measures intended to guarantee distant or close security cannot shelter all the elements of the Large Unit from the effects of distant artillery fire, gas attacks, or even completely from incursions by armored vehicles, etc.

It is the duty of the commander of any formation, whatever the place it occupies in the disposition on the march, on station, or in combat, to provide permanent measures for the local security of the element placed under his command.

These measures are of two kinds:

Active measures with a goal of:

— monitoring the terrain and the atmosphere;

— action by fire using appropriate weapons (antitank weapons).

The deployment of these weapons must, for them to be effective, be carried out as soon as possible and carefully prepared in all situations.

Passive measures include:

— dispersal of units and dilution of formations;

— use of camouflage, shelters, obstacles, and barricades;

— collective and individual protection against gas;

— precautions to be taken in stationing and deploying units.

The details of the prescriptions concerning local security measures are given by the regulations of the arms.

ARTICLE 3.

AIR SECURITY.

151. Air security aims to protect Large Units, on station, on the march, or in combat, against the enemy's air activities (aerial intelligence investigations, bombings, and ground attacks by heavy and light aircraft, airdrops, etc.).

This protection is ensured using two kinds of measures:

— some are the responsibility of the command, and together they constitute *air cover;*

— others, taken by all units, whatever their importance and place in the system, are called *local air security measures.*

152. Air security is based in particular on:

— *on intelligence;*

— *on the deployment* of light aviation and antiaircraft defense forces.

I. Air Cover.

153. Air cover can be provided when:

— the commander is *informed* in time of the possibility of an enemy air enterprise;

— his Large Unit is covered by the combined action of light aviation and antiaircraft defense forces.

154. *Intelligence* is provided concurrently with aviation by the watch system organized by the Large Unit in liaison with the general security units (lookouts, intelligence, signal communications) of the territorial antiaircraft defense (D.A.T.) established in the zone of action of the Large Unit.

Air cover is ensured by using a disposition combining the action of defensive light aviation, the D.C.A., and the D.A.T. under the conditions laid down in Nos. 301 and following.

155. The command, at all echelons, must also take measures to protect the troops, especially the sensitive points behind the front, against surprises resulting from *airdrops.*

The organization to be provided for this purpose must involve:

— territorial antiaircraft defense (lookouts, warning, artillery, machine guns, etc.);

— antiaircraft defense (artillery, machine guns);

— defensive light aviation.

Units of all arms, the closest to the landing site, preferably mechanized or motorized, will have the permanent obligation to intervene without delay against the enemy elements.

II. Local air security.

156. The security of units against the observation and fire of enemy air forces rests on:

— intelligence provided by a lookout service, which must be organized in each unit, on the march, on station, or in combat, in conjunction, whenever possible, with the air surveillance and warning system of the D.A.T.;

— a set of measures:

— *active* ones (rapid implementation of organic means of fire against aircraft flying at low altitude);

— the other *passive* ones aimed at shielding the troops from aerial observation (night movements, concealed and covered routes, camouflage, extinguishing lights, dispersal of units, and dilution of formations) and protecting them against fire (saps and shelters).

Passive measures are only taken, however, if the speed of the movement does not take precedence over the concern for protection.

The details of the prescriptions concerning the local air security measures are given by the regulations of the arms.

TITLE IV.

TRANSPORT. — MOVEMENTS. ENCAMPMENT.

CHAPTER ONE.

GENERAL INFORMATION ON DISPLACEMENTS.

157. Wartime operations require constant displacement of troops and supplies of all kinds.

Troop displacements take the form of *movements* or *transports.*

There is a *movement* when troops move by their own means, whether these belong to them organically or have been placed temporarily at their disposal. The movement is organized and regulated by the commander of each Large Unit according to the maneuver and within the framework of the Instructions given by the higher echelon.

Transport occurs when troops are moved by transport services independent of the units being transported. Transport is generally organized and regulated, in accordance with command instructions, by the transport service to which it is entrusted.

The displacement of a Large Unit may involve both movement and transport simultaneously; its motorized elements, for example, moving by their own means, while its foot and horse-drawn elements are transported by rail.

158. To obtain the best possible efficiency from transportation assets, in particular motorized means, and to thus exploit the continuous progress of science with regard to them, the displacements must be the subject, on the part of the command, of an organization and particularly of a judicious coordination of movements and transport.

159. The means of travel used are: normally railroads and roads, possibly inland waterways, sea routes, and, for light elements, air routes.

Standard-gauge *railroads* constitute for the high command a maneuver tool, *flexible* and *powerful,* the intensive use of which is most often essential to transport the Large Units close to their zone of use and to ensure the necessary transport for the maintenance and the deployment of their manpower and their means of action.

Roads make it possible, with the help of a dense road network, organized and suitable for intensive traffic, to access wherever the needs of the armies require it, thus extending the railroad line and making up for its deficiencies or its failures.

Roads have taken on capital importance in strategic transport due to motorization.

160. Regardless of the mode of travel adopted:

— there is always an interest in not breaking the organic links between the units;

— it is important to conceal from aerial observation and from enemy espionage any movement or transport of troops carried out by night or day. *Camouflage* discipline must be a primary concern of command. For this purpose, extensive use is made of night movements (No. 476).

— it is also necessary to maintain secrecy both on the route and on the point of destination of the Large Units during transport or movement.

— finally, all precautions must be taken to guarantee, depending on the situation and in accordance with the requirements of Title III, the *security* of movements and transport.

CHAPTER II.

MOVEMENTS.

ARTICLE 1.

GENERAL.

161. Despite the development of means of transport, the situation frequently imposes on troops the obligation to cover long distances by their own means.

Road movements have become complex; indeed, the elements entering into the composition of the Large Units and the General Reserves can include elements on foot, cyclists elements, elements on horseback and horse-drawn vehicles, and motorized elements, that have neither the same speed nor the same movement capacity. In addition, certain routes in the zone of movement of a Large Unit may be encumbered with constraints that limit their use by this Large Unit.

Road movements must therefore be carefully prepared and regulated at all echelons.

162. To this end, it is essential:

— to adopt marching dispositions appropriate to the different constituent types of the Large Units;

— to use a marching pace in keeping with the normal stages of these different types and the constraints of the road network.

163. When the tactical circumstances permit, Large Units should be split into *march groupings* comprising elements of the same speed moving on specialized or non-specialized routes.

These groupings must nevertheless meet the general *tactical conditions* such as those imposed by the disposition to be carried out at the end of the march, the order of priority of arrival of the elements, their installation, etc., and the conditions of a *technical order* that depend on the characteristics of the road network and the ability to move the various elements. Each of these groupings is placed under the leadership of a commander responsible for the formation, departure, progression, and local security of the column.

Taking into account the tactical factors, the assignment of the routes included in the zone of action of a Large Unit is regulated in such a way as to ensure as much as possible to each march grouping the independence of its movements, to the troops on foot and horse-drawn the minimum fatigue, and mechanized motors normal use.

The conditions for carrying out the movement of a Large Unit are fixed by the *movement order* of this Large Unit, which specifies, in particular, the composition of each march grouping, the route assigned to it and designates its leader.

164. The preparation of complex movements by land requires great precision. It is facilitated by the use of *march diagrams* to control at any time and to vary, if necessary, the movement.

165. The disposition of the march, the organization of the columns, their deployment, the choice of routes vary essentially according to the distance from the enemy and according to whether the troops carry out their movement *on roads* or *across fields.*

ARTICLE 2.

ROAD MARCHES.

1. Before establishing contact.

166. In open ground, before establishing contact or close to their zone of intervention in the battle, the Large Units move on roads and, in general, by their own means.

The command assigns each Large Unit its own *zone of movement.*

167. To facilitate the movement, the command, taking into account the traffic rules enacted by the superior authority, must, as far as possible, comply with the following prescriptions:

— let each element have its own character; for this purpose, form the columns as provided for in No. 163:

— assign the shortest routes to foot and horse-drawn elements;

— reserve paved one-way routes for heavy matériel:

— place, under the same command, various march groupings or take the same itinerary or the same zone of movement, each time that their echelonment does not place them at least at a staging distance from one another.

168. The Commanding General of the Army distributes the resources of the road network at his disposal between the Army Corps, according to their composition and their mission. He regulates the movement of the elements of the army or distributes them, for the march, between army corps.

Automotive elements generally move in large bounds, every two or three days.

The Army commander decides, according to the airfields, the conditions of movement of all the aeronautical units belonging to the army and to the Large Units that enter into his composition. These movements take place in bounds of at least 40 kilometers. In the event that, due to a lack of sufficient terrain, it is not possible to bring all the units forward, the Commanding General of the Army sets the order of priority of movements.

169. The Army Corps marches in principle by divisions placed abreast; however, an Army Corps with three to four divisions will generally maintain one or two in the second echelon. The organic elements of the Army Corps and the Army elements assigned to it for the march are grouped in the wake of a division or distributed among the divisions. The marching disposition is set in such a way as to leave sufficient independence to the subordinate units and to allow possible changes of direction.

When two divisions placed one behind the other use the same road network to march to the enemy, it is necessary to regulate their movement carefully, taking into account the *degree of priority* of the arrival of their various means on the battlefield; thus it may be advantageous to advance the second-echelon division with its fighting elements, the artillery in the lead, and to send all the cumbersome formations of the first-echelon division to the rear.

170. The division preferably marches in several columns, making best use of the zones or routes assigned to it.

II. Sheltered by a formed front.

171. Sheltered by a formed front, and in particular in the various staging zones, the Large Units on the move most often take routes likely to be used for other movements and transport. The time conditions under which these routes are made available to each Large Unit are set by the authority responsible for regulating movement and transport on the road network in question (road regulatory commission).

172. Distant from the front, in security, efforts are above all made to reduce the fatigue of the troops. For this purpose, the columns are organized under the general conditions provided for in No. 163.

A large enemy air force will often force them to march at night or to take air cover measures during the day.

III. Movements of Motorized Large Units.

173. Motorized Large Units are most often displaced entirely by road, with the assistance of General Reserve transport units placed temporarily at their disposal.

The movement is regulated by the commander of the Large Unit to whom the command assigns, either a zone of movement, or specific routes, when it is a question of movements sheltered from a formed front. In this last case, the movement order of the Large Unit takes into account, if necessary, the constraints that weigh on these routes and which are prescribed by the road regulatory commission established in the corresponding zone.

Motorized Large Units in march order are characterized by the considerable length of their columns.

In order to ensure the unloading of their elements and their deployment in good conditions of order and speed, these Large Units must have several motorized routes. This necessity often leads the command to assign them larger zones of movement than the normal type of Large Units.

The order of priority of arrival of the various elements is always the determining factor in the constitution of the columns.

Title X (No. 455 and 458) indicates the general conditions under which the movements of motorized Large Units are organized, taking into account the situation. It specifies in particular that the movement of their main body can only take place under the shelter of a solidly constituted front or inside the security polygon created by a set of security units (Large Cavalry Units, reconnaissance battalions, advance guards and flank guards).

ARTICLE 3.

CROSS-COUNTRY MARCHES.

174. As soon as the frequency and intensity of aerial bombardments or artillery fire, incursions by armored vehicles compel them to do so, the Large Units abandon road formations and continue their advance across fields.

They then take an approach march disposition under the conditions which are the subject of Nos. 209, 339, and 372.

CHAPTER III.

TRANSPORT.

175. The transport of troops and supplies is normally carried out behind the shelter of a front that absolutely guarantees their security.

ARTICLE 1.

RAIL TRANSPORT.

176. Rail transport is applicable to all arms. It is of considerable yield and lends itself to a certain number of variants.

However, its execution, which generally requires significant delays, requires preparation. Also, transporting a Large Unit by rail is only justified for movements with a range of at least 100 kilometers for an Army Corps and 75 kilometers for a Division.

It is essential for long journeys, but, following the development of motorization, it can happily be combined inside a Large Unit with the road movements of motorized elements.

The deployment thus achieved significantly reduces the transport times of the Large Unit.

Rail transport is difficult to hide from aerial investigations. Their protection against air attacks is ensured, on the one hand by the command by using antiaircraft defense and defensive light aviation, on the other hand by the organic means of defense of the units.

177. Rail transport is the most capable of supplying the armies with the power necessary in operations involving large numbers of troops. The terminus of the railroad lines must therefore be kept as close to the troops as the need to carry out unloadings and transshipments with sufficient security permits. The increased range of artillery, as well as the increased power of aviation, commands the dispersion and the deep echelonment of unloadings and transfers. The game of these operations must, however, have the necessary flexibility to adapt to the tactical and technical conditions of the moment.

The development of motorization at all echelons also offers increased possibilities in this area.

ARTICLE 2.

MOTORIZED TRANSPORT.

178. Motorized transport, likely to use most of the road network, is characterized by its great flexibility. The richness of this network lends itself to variants and allows motorized transport units:

— to unload the troops fairly close to their place of use;

— bring supplies close to interested parties. However, the performance of motorized transport depends essentially on the characteristics and the state of maintenance of the roads used as well as on the organization of traffic.

In general, it is less difficult to mask than rail transport. Motorized transport, however, has the disadvantage of being cumbersome and requiring strict regulation of traffic. Additionally, when carried out over long routes, they risk breaking up the large units transported when these do not have trains and motorized convoys.

179. It is possible to use motorized transport for most of the combat elements of the Large Units and the General Reserves, which are not themselves equipped with organic motorized means. However, this mode of transport has low yield when it comes to animals; moreover, the current limitation of the tonnage and size of motor vehicles does not make it possible to remove certain heavy or bulky horse-drawn matériel by truck.

180. Road transport can either extend rail transport or be combined with it for the movement of the same unit. The combined use of rail and road for transport of a certain amplitude is becoming more and more necessary, because of the development of motorization.

ARTICLE 3.

WATER TRANSPORT.

181. *Transport by inland waterway,* reserved in principle for the supply of heavy matériel and perishable foodstuffs and for medical evacuations, can bring appreciable assistance to rail and road transport, but it is too slow for the movement of troops, except in the case of major river arteries.

Maritime transport either connects theaters of operations where land connections are non-existent or insufficient, or else, in the same theater of operations, relieves the other modes of transport by participating in bypass transport. Questions relating to the transport itself are regulated by the naval authorities, in agreement with the military authorities.

ARTICLE 4.

AIR TRANSPORT

182. *Air routes* are normally used by the flying echelons of air forces.

It can be considered for isolated individuals, small detachments, supplies or medical evacuations.

Its importance is constantly growing and future prospects justify considering its use as less exceptional (No. 300).

CHAPTER IV.

MOVEMENT AND TRANSPORT COORDINATION.

ARTICLE 1.

GENERAL.

183. The speed of execution of the maneuver requires the use of all the resources of the signal communications network and, consequently, for movements of a certain amplitude, the combined implementation of rail transport and motorized movements and transport. Similarly, the development of motorization inside Large Units of all types leads the command to prescribe their movements both by rail and by motor vehicle.

184. The need to ensure at the end of the transport the regrouping of the elements of the same unit moved by different means, that of combining, taking into account the situation, the use of these means under the best conditions, make it essential close co-ordination of rail transport and motorized movement and transport.

185. Moreover, the same road network is often used by diverse movements and road transport. To obtain the best performance from this network and avoid any traffic hindrance to the various elements, it is important that in a zone or on determined routes, road movements and transport are organized and regulated by the same authority.

ARTICLE 2.

COORDINATION OF RAIL TRANSPORT AND MOTORIZED MOVEMENT AND TRANSPORT.

186. The coordination, in each particular case, of rail transport and major movements and motorized transport is ensured, at an initial stage, by the instructions that prescribe the use of these two modes of transport. These instructions originate, in principle, from the commander-in-chief, to whom the railroad service reports directly. They define the portion allocated to the railroad and the portion designated for the road. They establish the conditions for organizing the trip and specify the authorities or entities responsible for its execution, as well as the regrouping of units at the end of the trip.

187. This coordination is ensured, at a second stage, within the framework of the army receiving the displaced elements. In fact, it is generally up to the army commander to set the final destinations of these elements. The transport services must, therefore, ensure the end of the route in accordance with its directives. For this purpose, the commander-in-chief seconded to the army commander a qualified representative who gave the necessary instructions both to the railroad service and to the road transport service.

188. In certain special cases (transports carried out in the zone of the same army), the army commander may receive a delegation from the commander-in-chief for the use of the railroad. The coordination between the ground transports and the road transports or movements with which they are combined is then entirely ensured by the army commander. Rail transport is regulated, on his instructions, by the representative of the military service of the railroads who is normally responsible for serving the army. The latter receives from the commander-in-chief, if necessary, the necessary delegations of powers and from the army commander the details relating to the elements to be transported by rail as well as their destination.

ARTICLE 3.

COORDINATION OF ROAD MOVEMENTS AND TRANSPORT.

189. Both in the area of the forward Large Units and in the various staging areas, the coordination of road movements and transport requires at each level:

— *the organization of traffic;*

— *the centralization of movements and transport* by a specific authority or body.

With regard to the forward Large Units in the process of movement before establishing contact, the movement order of the Large Unit most often replaces the plan of movements and transport (No. 191).

Traffic organization.

190. The Large Units have the road network included in their zone of movement or in their zone of action, except for constraints weighing on certain routes that the higher echelon intends to use for the displacements regulated by its care.

At each echelon, the command has the duty to take responsibility for the organization and monitoring of traffic on the roads of general interest. This action is exercised there through specialized bodies: road regulatory commissions and road traffic detachments.

When the road network is sufficiently dense, particularly in the forward zone, there is no point in regulating traffic on all the routes.

Traffic regulations are the subject of *the traffic plan.* This document, supplemented by a sketch, is more or less permanent. It must in particular:

— set general and local instructions;

— identify the roads that will be subject to regulations: dispatch routes or likely to become dispatch routes upon receipt of notice in order to deal with certain contingencies;

— define the priority to be granted on the various routes;

— distribute traffic surveillance personnel.

Centralization of movements and transport.

191. The coordination of movements and transport is the subject of a *movement and transport plan in each Large Unit.*

This document applies to a determined period (in principle 24 hours); it regulates the movements and transports to be carried out firmly and also makes it possible to organize unexpected movements and transports quickly and judiciously.

In particular, it must identify:

— the conditions under which the regulated network must be used by troops and transport units: assigned routes, entry and exit points, time, and possibly speed of march;

— the distribution of the transport to be carried out between the various means: horse-drawn train, motorized train, narrow gauge railroad;

— the terms and conditions for using the tonnage appropriation allocated to the Large Unit on General Reserve motorized transport units.

The traffic units (road regulatory commissions and road traffic detachments) receive communication of this plan and ensure regulation.

CHAPTER V.

ENCAMPMENT.

192. Under the pain of ruining the troops, it is essential to give them time to rest, prepare their food, look after their animals, and maintain their matériel.

However, the requirements of the situation take precedence over the concerns of convenience and comfort and impose the choice of various parking methods.

193. The encampment order assigns a cantonment zone to each Large Unit or important grouping (organic elements of an army corps, convoy groups, etc.). This area is divided at each echelon between the subordinate elements. Troops move in, trying to break tactical ties as little as possible.

194. Both in camp and on the march, meticulous precautions are taken to conceal large gatherings from aerial observation, both by night and by day.

The units must liaise with the territorial antiaircraft defense units (lookouts, intelligence, signal communications) located nearby or improvise, by their own means, a lookout and warning service.

At night, regardless of the encampment mode, lights and traffic lights must be camouflaged and even turned off.

Avoid accumulating troops in tight spaces.

195. Encamped troops must be able to be removed with minimum delay. Their removal is prepared by designated representatives of the transport services who enter into liaison with the commander of the troops for this purpose.

Cantonments.

196. Far from the enemy, the troops are quartered either in the *cantonment zones* that coincide in principle with the zones of movement assigned to the Large Units, or in the rest areas organized behind a formed front.

They are distributed there according to the best of the situation and the resources offered by the inhabited places or the camps.

Columns are most often stationed along and near the march route to a depth corresponding to the length of the column. They thus rest in the best conditions, without being obliged to add lateral movements to the stage covered. In addition, all the elements of the column being able to arrive at the resting place and to leave it almost simultaneously, it is possible to use all the capacity of the troop.

Cantonment. — Bivouac.

197. When approaching the enemy, the decrease in the column length leads to a corresponding decrease in the depth of the cantonment zones.

The tightening of the cantonment zones progressively takes place to avoid excessive fatigue on the rear units of large columns.

The troops take shelter in villages or bivouacs nearby in order to exploit the resources they present. The threat of aerial bombardments will frequently make it necessary to give up occupying the villages using their maximum density.

The troops then station in *cantonment-bivouac.*

Bivouac.

198. *Near the enemy,* in the zone exposed to long-distance enemy artillery fire and frequent aerial bombardment, the troops are encamped in very open formations; most bivouac.

The bivouacs are chosen in such a way as to escape enemy ground and air observation and to allow the troops to assemble and move.

Tactical links are respected in order to facilitate the regrouping of Large Units.

Avoid bivouacking near points likely to be systematically chosen as targets by enemy artillery and aviation.

In the immediate vicinity of the enemy, troops are often obliged to station themselves in approach or even combat formation. They use woods, cover, existing shelters, or dig in by camouflaging their work.

Security of cantonments and bivouacs.

199. The local security of the cantonments and bivouacs is ensured by the execution of the measures provided for in Nos. 150 and 156.

Among these measures, protection against gas, incursions by armored vehicles, and airdrops must retain the particular attention of the local command.

TITLE V.

THE BATTLE.

CHAPTER ONE.

FEATURES OF THE BATTLE.

200. The object of battle is to break the matériel power and the psychological force of the enemy. On the *offensive,* it drives him out of his position, breaks his disposition, and continues the destruction of his forces; *defensively,* it makes it possible to hold ground by repelling the opponent's attacks.

By conducting an offensive battle, the command is often led, for the purpose of maneuver, to prescribe defensive actions on certain parts of the front.

201. *The offensive battle* presents, at least at its beginning, different features depending on whether it is the culmination of a maneuver or whether it is fought against a stabilized front.

Three possibilities are to be considered:

— *the two adversaries are on the offensive and are continuing their advance.* The establishment of contact and the intervention of the first-line units then follow one another quickly and can sometimes lead, in a short time, to a thorough attack and a decision. The one of the two adversaries who has been able to adjust his maneuver in such a way as to carry out the encounter on a *battlefield* of his choice will be placed in the best conditions for success; or

— *the enemy on the offensive march halts to resist,* protects himself with entrenchments, and adjusts his infantry and artillery fires. The rhythm of the operations to be undertaken by the assailant will be more or less accelerated according to the degree of advancement in the deployment of his disposition, his ammunition supplies, and the strength of the enemy's organization; or

— *the enemy presents a fortified or stabilized front.* A rigorous method is then essential in the attack, which requires long-term preparations intended to bring on site the necessary means.

In general, especially at the beginning of a war, and while reserving the share of legitimate initiatives, it is important to fight **directed battles** ***and to avoid*** **encounter battles.** ***These, in fact, by the hazards they entail, do not lend themselves well to the use of young troops who, on the other hand, must only be engaged methodically on the battlefields, with all the necessary fire support.***

202. *When the enemy's front is broken,* the victorious troops attach themselves, without letting go, to the defeated enemy and press him without delay to prevent him from reconstituting. When they come up against organized resistance again, they make close contact with it as soon as possible, and the command gathers the means to proceed with a new attack.

The *offensive* battle, therefore, takes the form of actions of successive forces, preceded by periods of halting essential for their preparation and followed by more or less long periods of movement.

203. *When the defeated enemy is no longer in a condition to resist* and abandons the fight, the pursuit begins. The undisputed occupation of the battlefield is not sufficient, but it is necessary, to complete the disorganization of the enemy and prevent him from reconstituting his forces.

If the attacks have not succeeded, the command seeks to limit the consequences of the failure by securing possession of the occupied terrain, at least for the time necessary to restore order to its disposition.

204. To fight a *defensive battle,* the command determines a position of resistance on which it fights the battle with all its means. This position is advantageously established, sheltered by an antitank obstacle. It is covered, in principle, by a system of more or less dense outposts and variable missions depending on the case.

If the enemy penetrates the position of resistance, he is thrown back by counterattacks; if, despite these counterattacks, the enemy succeeds in seizing this position, the fight is resumed on the rear positions, which it is up to the *higher command alone,* in anticipation of such a risk, to have organized and occupied in proper time.

The defensive battle can also be only the preparatory phase of an offensive maneuver decided in advance. In this case, when the command considers the assailant sufficiently worn out, it orders the start of the *counter-offensive;* the movement is started by reserves of all arms, with the support of all the artillery of the line of battle.

205. Whether offensive or defensive, the battle is often characterized by the *duration* and always by the *rapid attrition* of the troops.

The duration is a function of various elements, particularly the power of the armament, the solidity of the organization of the terrain, the relationship of the forces present, and the physical and, above all, psychological state of the two adversaries.

Attrition is brought on by the emotions of the struggle, the losses suffered, and the fatigues endured.

These characteristics mean that to sustain the effort required in battle, commanders need a large number of reserves, both to relieve the troops engaged and to allow for maneuvering combinations. Refilling these reserves must be a constant concern.

CHAPTER II.

THE OFFENSIVE BATTLE.

206. *Unity* in action is the essential guarantee of success in the offensive battle.

The leader will obtain this necessary coordination of wills, initiatives, and efforts by giving each Large Unit or subordinate unit:

— *a mission;*

— *a direction,* resulting from that which he will himself have received from the superior authority.

The direction is imperative. It is the basis of strategic and tactical discipline.

— *objectives* to be achieved by the main bodies carrying out the maneuver of the Large Unit or allow the reorganization of its disposition.

207. Any offensive battle involves an echelonment of efforts over time that *can be broken down as follows:*

— *a preliminary phase* aimed at:

— conveying the means in the direction of the enemy, **in security** and with few losses;

— clarifying, **through combat,** the adversary's situation;

— constituting a **front** under the shelter of which the means can be, under favorable conditions, arranged for the attack.

— an *execution phase* involving the application of the main body of the forces in a chosen direction.

— *an exploitation phase* that aims to complete the dislocation of the system of enemy forces.

The preliminary phase, which includes the *approach march,* the *establishment of contact,* and the *engagement,* is extended more or less according to the time that is necessary to establish contact, to identify its value, to fix the enemy, and to establish a front.

The execution phase, an actual act of force, is characterized by power and violence developed to the highest degree, by throwing all available forces into the attacks if necessary.

Exploitation of success is undertaken without delay so that the enemy does not have time to mitigate its effects.

208. *If it seemed necessary, in order to classify ideas and analyze them, to divide the battle into distinct sections, it is important to emphasize that these different phases do not always follow one another, in reality, according to this immutable order.*

The general rules that are exposed, during the articles hereafter, about each of the phases of the battle, have thus for an object only to guide the leader in the choice of the procedures that he will have to adapt to the circumstances and, in particular, to the mission assigned to him.

ARTICLE 1.

APPROACH MARCH.

209. Approach marches begin as soon as the intensity and frequency of aerial bombardments, artillery fire, or incursions by armored vehicles force the Large Units to abandon road formations.

They then take so-called approach march formations that aim to carry them towards the enemy, in secret, safely, as quickly as possible, and in the best physical and psychological conditions.

For this purpose, the command strives to control enemy air force, artillery, and armored vehicles or, at the very least, to protect the troops against their effects.

210. *Night,* being unfavorable to the effectiveness of the adverse actions and conducive, on the other hand, to the preservation of secrecy, will be frequently used for the approach march, despite the fatigues that it imposes on the troops. The night approach march is especially prescribed when the advance is carried out under the shelter of a formed front or in the direction of an adversary in position; in the latter case, effective distant security is essential.

Circumstances will often require, however, the carrying out of the approach march *during the day,* especially when contact appears to be imminent. The troops, leaving the main roads particularly under observation or beaten by enemy fire, then use the paths and tracks and, if necessary, march across fields, taking routes as concealed as possible from ground observation posts and aerial observation.

211. The approach march disposition of the Large Units must, as far as possible, contain the seeds of, at all times, the projected maneuver and comprise a sufficiently flexible deployment to lend itself to the successive transformations imposed by the circumstances.

It generally includes:

— in the first echelon, the Large Units whose mission is to establish contact;

— in the second echelon, the Large Units intended for the maneuver. The artillery of the Large Units is arranged in such a way as to intervene quickly in the action when the moment comes, taking into account the double need to guarantee its security and not to delay the entry into line of the infantry.

All non-essential items on the front are kept in the rear.

In a Large Unit located on a wing, the disposition must be able to lend itself to the protection of the open flank, to the extension of the front, or even to the execution of an enveloping maneuver.

212. The disposition must be constantly adapted to the current situation. When far from the enemy, it is broad and deep to limit the troops' fatigue, make the best use of the road network, facilitate changes of direction, an enveloping maneuver, or protect the flanks. The disposition tightens as it approaches the enemy. Thus, the most distant units can intervene in good time.

213. The disposition moves in bounds, echeloned in depth, and split into march groupings.

The command regulates the bounds of the main bodies according to the successive battlefields it envisaged in its forecasts, taking into account, if necessary, constraints inherent in the movement of the motorized formations.

The bounds of the security units are themselves determined according to those of the main bodies and in such a way as to best screen the latter against the interventions of the adversary's motorized or mechanized means.

214. The disposition's protection against armored vehicles is based on the allocation to the advance guards of antitank weapons ready to intervene without delay and even of a few tank units more specifically intended to stop enemy vehicles as distant as possible.

There will often be an interest, moreover, in keeping Large Units on the march behind major terrain breaks in order to oppose the incursions of enemy armored vehicles into the rear of the zone of movement.

Air cover of the entire disposition is organized under the conditions provided for in No. 302.

215. If, during the maneuver, certain Large Units are led to deviate from the direction that has been assigned to them, the command brings them back gradually or takes all useful measures, such as the introduction of a second-echelon Large Unit, to keep the whole disposition on its direction.

ARTICLE 2.

ESTABLISHMENT OF CONTACT.

216. The approach march results in contact, which aims to:

— in front of an adversary in position, to identify the line on which he offers organized resistance;

— in front of a moving adversary, to stop his progress on terrain chosen by the leader and to fix the enemy there before attacking him.

In both cases, ***the establishment of contact tends to create a front behind which the main bodies will complete their preparations for the attack.***

217. Establishing contact is the progressive work of all the elements preceding the main bodies and particularly of the advance guards (No. 144).

The commanders of the Large Units endeavor to attenuate the slowness of the advance by the impetus that they give to it, by the flexibility of the command organization, and by a liaison of the arms allowing fast and effective support.

218. *In front of an enemy in position,* the advance guards operate without worrying about alignment, first by infiltration, then by more or less isolated maneuvers and combinations of fires, soon welded and completed by neighboring actions. They finally engage all their means to make deep contact and create a resistant front.

Thanks to their protection, speed, and armament, armored vehicles promote and accelerate contact and, on the other hand, make it possible to save infantry. In this respect, the allocation of tanks to the security units is often justified, subject to avoiding their systematic distribution between the elementary units.

219. *In front of an enemy on an offensive march,* the advance guards, after having advanced under the cover of distant security elements, set up defensively, as soon as the order is given, on the position where the leader has decided to let the opponent come into contact.

This position is chosen because of its defense facilities, particularly against armored vehicles, and because of the benefit it presents for the resumption of the offensive movement.

The order to deploy to this position is given at the appropriate time to allow the enemy advance only the slightest gains in terrain and yet allow the advance guards sufficient time to adapt a coherent fire system.

The defensive installation is carried out under the protection of distant security units and in the minimum amount of time. All available artillery is deployed to participate in this blocking action.

220. During the initial contact, each echelon of command monitors the operations of the security bodies very closely. He takes special care of his permanent mission of intelligence in order to allow the leader to develop his plan of attack with full knowledge of the facts.

Aviation must help to shed light on the situation by seeking beyond the line of contact for significant enemy forces, as well as any clues likely to enable the command to assess the adversary's possibilities.

ARTICLE 3.

ENGAGEMENT.

221. The results of the establishment of contact will generally not allow an attack to be made without halting.

This will be the case, in particular:

— when the intelligence obtained by the establishment of contact needs to be clarified or supplemented;

— when the enemy has retained important observation posts or salient points of the terrain, the contact front will not provide the possibility of proceeding, under good conditions, the assembling of forces for the attack;

— when the security units have not been able to seize the outposts of an adversary in position, the attack would risk going off course and too far.

In such circumstances, the command is required to prescribe preliminary actions, the whole of which constitutes the *engagement.*[1]

Their purpose is to identify the value of the contact and to seize the positions or strong points necessary for the deployment of the attack disposition and its debouchment.

In order to maintain the high levels of availability required by the development of the battle, it is important to provide the first-echelon Large Units in charge of engagement actions with very energetic matériel support: artillery and tanks.

As opposed to the decentralized actions of the approach march and the establishment of contact, the command must, therefore, intervene during this phase to coordinate the efforts with a view to the subsequent attack. As such, the engagement *must be considered as the first act of the attack.*

Within the overall framework of the future operation it has designed, the command specifies the points at which it intends to test the enemy's force and the possession of which will make it possible to cover the deployment of its attack disposition.

[1] This is similar to the phase of the operation that the U.S. Army commonly referred to as a *developing attack* or a *reconnaissance in force.*—Trans.

It can also define the *engagement front of* the units called upon to engage.

ARTICLE 4.

ATTACKS.

222. Whatever the general form of an offensive maneuver, there always comes a time when the enemy manages to oppose a front that must be broken by force.

The attack is the act that characterizes the offensive battle.

The ***concentration of efforts*** obtained by the centralization of command is its distinctive feature.

223. Against an enemy on guard, the deployment of forces for the attack is prepared with methodical precision.

Against an adversary who is incompletely established in positions known to the assailant, the search for sudden and brutal action, on the other hand, takes precedence over the advantages of meticulous preparation.

The benefit of sudden attacks is as considerable if the enemy is surprised as the risks are great if he is warned.

224. The greater the initial size of the attacking front, the greater the results that can be expected from an attack. It is through the flanks, which are subjected to converging fire, that an attack is most often slowed down; its front thus gradually narrows and, if it does not start from a sufficiently wide base, it is extinguished before it has achieved any exploitable results.

225. An attack front can include passive parts provided that the attacks are combined and welded by the fires and joined during the advance.

These attacks will not all be of equal importance.

One of them, responsible for the main effort and called the main attack[1], will be carried out with all the means that the leader can apply in the direction where success will give decisive results. The others will be used to frame or cover it.

From the point of view of execution, there is no distinction to be made between these various attacks; they will all be pushed to the limit without worrying about alignment.

226. The possibilities of the command in means of all kinds closely condition the extent of the attack fronts that it can envisage. They will,

---

[1] The terms main attack or main effort shall not, as a general rule, appear in orders at the division echelon and below this Large Unit.

therefore, necessarily have their influence on the depth of the possible objectives and consequently on the choice of the most profitable directions.

But we must not forget that these possibilities depend to a very large extent on the energy and skill with which the leader will be able to practice ***economy of forces*** and push it to the last degree. Often, moreover, he will be able to overcome the undeniable risks thus incurred by the very direction given to the main attack, which immediately threatens the adversary in his sensitive points and will impose a defensive attitude on him and keep him there.

I. General disposition.

227. The attack cannot generally be carried out with only the organic means of the Large Units. It requires the allocation of additional means that almost always include artillery units, tanks, and sometimes machine gun battalions.

Tanks are particularly adept at helping infantry overcome the difficulties they encounter during the advance.

Machine gun battalions can hold the passive parts of the front to allow for stronger concentrations of force on the active parts. They can also, at the debouchment of the attack, reinforce the bases of fire, and protect the flanks of the advancing units.

The command distributes these additional means. It first provides for the needs of the subordinate units and then reserves the use of the remaining means.

228. A Large Unit's attack disposition generally includes:

— in the first echelon, units placed abreast, deployed facing their objectives, in zones of width corresponding to the importance that the commander attributes to their effort;

— in the second echelon, reserve units intended either to relieve, overtake, or reinforce the first-echelon units, or to deal with the unforeseen events of the battle.

229. The artillery, subject to being covered in particular against armored vehicles, is deployed as far forward as possible so as to reduce the number of its movements during the operation.

230. The use of numerous tanks, spread over a wide front and echeloned in depth, is the rule for their use in the attack.

This employment produces a powerful psychological and matériel effect, promotes the advance of the infantry, and encourages enemy artillery and antitank weapons to disperse their fire.

The leader takes the initiative when the main mass of tanks enters into action. Sometimes, these will intervene at the start of the operation; sometimes, this intervention will only be considered after the conquest of one or more objectives.

Tanks can take advantage of their speed to advantageously take their initial position at a distance from the base of departure that puts them out of reach of enemy counter-preparations.

Tank units can be:

— either subordinated to the infantry for an accompanying mission.

Tanks of all models can receive such a mission; for the duration of the latter, they are said to be *accompanying tanks.*

— or maintained under the orders of the commander of the Large Unit when they intervene in battle for the benefit of the overall maneuver.

For the duration of their mission, these units are called *mass-maneuver tank units.*

In the first case, mixed infantry and tank groupings are formed; their command in breadth and depth is organized while avoiding, as far as possible, any dissociation from the organic links of the units.

In the second case, the tank units receive successive missions, precise and limited in space. The leader of the Large Unit then coordinates their action with those of the artillery and the other units of the general disposition.

231. The command takes all the necessary measures to protect, in all its depth, the offensive disposition of the Large Units against the attacks of enemy armored vehicles (No. 385).

II. Preparation for the attack.

232. The preparation of the attack is the set of fires executed before the H hour (No. 233).

It is mainly the work of the artillery, but the infantry and the air force can also participate.

The extent and duration of the preparation varies according to the circumstances. Sometimes, the preparation aims at the destruction and neutralization of the enemy's units in a deep zone; sometimes, it simply aims at the neutralization of the first opposing elements. ***Exceptionally,*** the attack is carried out without preparation. It is then said that it takes place by surprise.

The decision on these points depends on the enemy's organizations and the means available to the command. Thus, a large endowment of attack units in tanks and a large amount of artillery, capable of instantaneous and powerful effects, make it possible to reduce and, *exceptionally,* eliminate preparation.

III. Execution of the attack.

233. The debouchment of the attack must be controlled by the command without any ambiguity.

Zero hour, as a general rule, is when the first echelon elements of the attack disposition must debouch from the base of departure.

If attacking units are at different distances from the first enemy line and such that it is necessary to cover with artillery fire the advance of these units from their respective base of departures to the first enemy elements, the H hour can be that of the beginning of these artillery fires. The debouchment for each unit is then adjusted according to its particular situation.

The H hour is communicated in good time by secure methods. In improvised attacks, it is often replaced by a signal or a set of signals whose appearance triggers the start of the attack either immediately or after the lapse of a fraction of an hour, these conventions being moreover constantly renewed to ensure secrecy.

234. The command also sets, at least for deep attacks, the *rhythm,* that is to say, the conditions of the advance towards the various objectives: average speed of advance, duration of the halt on the objectives, conventions for resumption of movement from successive objectives, designation of the authority responsible for ordering it, movement of artillery, etc.

These arrangements are drawn up all the more minutely when the adversary is more powerful and has had more time to organize himself.

235. Whether or not there is artillery preparation, the attack debouches under the cover of an intense action of ***the entire mass of artillery***.

As a general rule, this action is exercised:

— by fires (*close-support fires*) covering as closely as possible the first elements of the attack (infantry and tanks) and intended to neutralize the resistance encountered successively during the advance;

— by fires (*protective fires*) which, extending the close-support fire into as deep a zone as possible, neutralizes the more distant points from which the enemy can act by the fire of his automatic and antitank weapons or which provide close views of the attack terrain;

— by a series of fires (*counterbattery and long-range action fires*) whose profound effects aim to control the opposing artillery, blind the distant observation posts, and oppose the action of reserves and supplies.

Close support and *protection* fires are provided by groupings:

— some called *direct-support groupings*, whose fires are driven directly and, by priority, by the first echelon of the attack disposition;

— the other so-called *general-action groupings*, acting in accordance with the orders of the leader of the Large Unit.

Counterbattery and *long-range action* fires are provided in principle by specific groupings.

236. Whether or not the attack is equipped with tanks and whatever the mission assigned to them, the action of the artillery takes place as it has just been said; however, the methods of this action must be adapted to the nature and characteristics of the elements forming the first echelon of the attack disposition.

When the attack has no tanks, or when tanks have only been given an accompanying mission, infantry units or, as the case may be, mixed infantry-tank accompanying groupings are supported by direct-support groupings. Immediate support fire is then carried out as close as possible to the first elements of the system, taking into account only the safety zones of the projectiles used and, if there are accompanying tanks, the clearance necessary for the maneuvering of these tanks (No. 393).

When the mixed infantry-support tank groupings are preceded by mass-maneuver tanks (No. 230), it is the latter that are supported by direct-support groupings. The close-support fires are then established far enough in front of the first echelons of mass-maneuver tanks to allow them largely to use their armament and their speed without fearing the projectiles of the friendly artillery.

In the latter case, all the close support and protection fire takes the form of successive box barrages (*encagements*)[1], types of enclosed fields, inside which the mass-maneuver tanks and then the mixed infantry-accompanying tank groupings advance by bounds, overpowering immediate resistance by their own means. The organization of the artillery command must then be planned in such a way that, in the event that the mixed infantry-accompanying tank groupings cease to be covered by the mass-maneuver tanks, each of these groupings can benefit without delay from close-support fire and protective fire.

237. In any case, the artillery stands ready to cover the infantry and the tanks during their halts on their objectives.

238. The air force deepens the artillery fire on the rear of the battlefield with its bombs. Sometimes, when its resources permit, it also precedes the first echelons of the attack disposition with its machine gun fire.

IV. Command action.

239. At all times of the battle, the commander of the Large Unit mainly makes his action felt: by the maneuver of the fires of his artillery, by

---

[1] *Encagement* (n) = Artillery fire aimed at isolating the opposing forces by cutting them off from their reinforcements and interdicting their supply.—Trans.

the intervention of the aviation (No. 296 and following), and especially by the engagement of his reserves (infantry and tanks).

He can, during the attack, call the artillery to exert, concurrently with the aviation, mass actions intended to deal quickly with the unforeseen events of the initial fire plan, either to obtain a reinforcement of the fires on a part of the front attack, to disperse enemy formations during assembly, to break a counterattack of infantry or armored vehicles, or to cover a threatened flank.

He regulates the deployment of infantry or tank reserves, gradually bringing them closer so as to ensure their timely intervention in the zone where he wants decisive results.

He engages them either to ensure the duration and continuity of the effort, to reinforce a Large Unit that is advancing, to restore in the desired direction a compromised maneuver, to fill a void, cover a flank, to limit a failure, etc.

As their engagement advances, the leader takes care to reconstitute his reserves by all the means in his power, but he does not hesitate to throw, if necessary, his supreme resources into the combat to achieve victory.

Constantly ensuring the coordination of the wills and initiatives of his subordinates, all of whom he has directed towards the common goal, the leader thus conducts the battle, gives it the character of unity essential to success, and deeply imprints on it the stamp of his personality, both in execution and design.

ARTICLE 5.

COMPLETION OF THE BATTLE.

240. If the attack succeeds, it is exploited without delay to complete the disorganization of the enemy and prevent him from reconstituting his forces.

Moreover, the command, most often, will not obtain such results by rushing its available resources through the breach created, which is generally too narrow. It should seek to widen the rupture zone by bringing down the uprights framing the breach.[1]

This widening can moreover be obtained either by rolling up the flanks, by progressively extending the attack zone, or finally by combining these two processes.

---

[1] This emphasis on the need to widen the breach is one of the main contrasts between the French tactical doctrine and that of the Germans, which called for maximizing the depth of the penetration in order to reach the enemy's artillery positions and to allow the envelopment of enemy positions. In the German system, it is the responsibility of reserve units to roll up the flanks of the newly created breach.—Trans.

241. Exploitation of success hampered by destruction, hampered by the rear guards, or carried out by troops already tested by combat risks being slow, progressive, and jerky.

One strives, at all costs, to maintain or regain contact with the adversary.

The Large Units are gradually moving towards a more and more decentralized and echeloned in depth disposition, split into combined arms groupings pushed vigorously forward and whose actions are coordinated by the indication of directions of exploitation targeting the weak points of the adversary: his gaps, his flanks and his lines of retreat.

As soon as the situation authorizes their use, the disorganization of the enemy is requested primarily from mechanized detachments (No. 38) and the air force, which attack the columns and convoys of the retreating enemy and endeavor to get ahead of its delaying elements on the terrain breaks. Cavalry units, and then motorized units, intervene progressively to preserve the result of these actions.

The rapid restoration of the communication routes is urgently undertaken.

If the enemy establishes itself in a new position, a new attack is organized as quickly as possible.

If the defeated enemy abandons the struggle and withdraws in disorder, the pursuit begins; it must be uninterrupted, daring, and relentless.

242. When the attack has not succeeded, the command first assures solid possession of the conquered ground; it proceeds, under the protection of the artillery, to put the disposition back in order, to withdraw the most tried units, and to their prompt reorganization.

It replenishes ammunition and manpower in order to be able to resume the offensive as soon as possible.

ARTICLE 6.

SIGNAL COMMUNICATIONS IN THE OFFENSIVE BATTLE.

243. *During approach marches*, the organization of signal communications is characterized by a system that must be flexible enough to be able to adapt quickly to the needs that will arise from the development of the maneuver. It must, in particular, ensure the signal communications of the intelligence bodies as soon as possible.

During the establishment of contact and the engagement, it also aims to achieve, in due time, the necessary liaisons to support the units responsible for these operations.

During these phases:

— the widest use is made of rapid signal communications agents;

— the establishment or extension of the telephone (or telegraph) network is prepared in the zones where the command has planned its effort.

Radio means are used only in case of absolute necessity (air defense, planes, cavalry units, etc.).

For the attack, the organization of the signal communications must allow the command to quickly activate artillery fires, aviation, as well as reserves.

Closely adapted to the situation of the moment, it must, however make it possible to deal quickly with new situations; its development varies with circumstances.

Against an adversary installed in a position, the delays necessary for the preparation of the attack and the halt times imposed in the advance facilitate the realization of a very thorough organization of signal communications.

Against an incompletely established enemy, the rapidity of the attack generally only allows the establishment of strictly indispensable signal communications.

During the preparation of the attack, the telephone network constitutes the backbone of the signal communications system; it must lend itself as well as possible to the rapid extension that the advance will require.

The radios are ready to go into action on command orders or in the event of an enemy attack.

During the attack, all the means are implemented without restrictions, in particular the radios, which will frequently be the only ones capable of permanently ensuring the necessary liaisons.

As the extension on the attack terrain of a developed telephone network presents considerable difficulties, it will generally be advisable to confine oneself to the relations with the most important subordinates and to carry out the other relations gradually by following an order of priority.

In the exploitation of success and in the pursuit, the organization of signal communications must, first of all, assure the command of close liaison with the units responsible for maintaining or resuming contact. It must, moreover, be able to adapt quickly to the various groupings formed in the system and then to the detachments responsible for initiating the pursuit.

Very wide use is made of rapid signal communications agents and radio methods.

ARTICLE 7.

SPECIAL CASES OF THE OFFENSIVE BATTLE.

I. Attack on a stabilized front.

244. On a stabilized front, the offensive battle takes place starting from close contact with the enemy's position of resistance, established for some time and controlled by preliminary partial operations intended to avoid surprises or even to improve the departure conditions of the attacks. It begins with the assault.

245. Knowledge of the enemy's organizations and the cover provided by the troops established on the front make it possible to undertake preparations for the attack well in advance. One strives to achieve them without giving warning to the enemy in such a way as to maintain the benefit of surprise.

The attack requires considerable matériel resources and, above all, the use of powerful, numerous, and well-supplied artillery, the firing of which will be carefully prepared.

It will be facilitated by the use of very powerful tanks.

The commander may have an interest in reserving the latter to act on the organizations that still hold in the depth of the enemy position while, due to the advance of the attack, the actions of the artillery become disunited and, therefore, less effective.

Surprise plays an important role in success; it will have to be sought even when extended artillery preparation is required, choosing the moment of the assault wisely.

Due to the development and the strength of the organizations that stabilization will have allowed, the battle will frequently last long enough to require the organization of successive attacks intended to shake up or break down the opposing front until the moment when a last push, taking advantage of the attrition of the enemy, will defeat him definitively.

II. Attack on a fortified front.

246. The attack on a fortified front is carried out according to the same principles as that on a stabilized front, but with even more powerful means.

CHAPTER III.

THE DEFENSIVE BATTLE.

247. The defensive attitude, the general characteristics of which have been defined in Title II, may take one of the following three forms, depending on the circumstances:

— *defense without thought of retreat,* which consists of holding a given position despite the enemy:

— *withdrawal from action*, a deliberate undertaking, the object of which is to offer the enemy successive resistances and to slip away from his contact in order to gain time or to lure him to a chosen terrain;

— *retreat,* which, following a failure, aims to withdraw the main body of the forces from enemy pressure under the protection of rear guards.

The first of these forms of defense must be considered as the rule, the others only the exception corresponding to particular circumstances.

ARTICLE 1.

DEFENSE WITHOUT THOUGHT OF RETREAT.

I. General characters.

248. To hold, in spite of the enemy, a given ground, with the defense organized on a position of resistance, covered by a natural or artificial ***obstacle:***

— initially, delays the adversary, as far as possible, by the action of advanced elements, long-range artillery fire, and by destruction;

— then endeavors, using its artillery and infantry fires, to break up the enemy's formations and to abort his attacks:

— finally, it concentrates on the assailant, at the moment when he is about to cross the obstacle, the fires of all its arms.

The fight continues, if necessary, inside the position of resistance and by the intervention of the reserves.

On this position, each defender must resist until the end and be killed on the spot rather than withdraw.

The strength of the defense is, therefore, based on a good system of fires combined with the obstacle, a judicious organization of the terrain, and the maneuvering of the reserves.

II. The defensive position.

249. A *defensive position* essentially comprises a *position of resistance* most often covered by a system of outposts.

The position of resistance has a depth defined by its outer edge or *main line of resistance* and by a *stop line*[1]*;* it being understood that these two terms do not express a linear arrangement of forces but simply represent a general outline.

250. The occupation of a defensive position is characterized by the ***echelonment in depth of the means,*** especially the means of fire inside the position.

This echelonment guarantees security, reduces the vulnerability of the troops, promotes the eventual re-establishment of the fire barrage, allows counterattacks to be set into action, and imposes successive efforts on the enemy. *It is emphasized all the more when the value of the obstacle is weaker and the organizations are less advanced.*

However, it cannot entail a uniform distribution of troops over numerous lines of defense, as this will result in a dispersal of forces incompatible with the solidity of the defensive system.

The position of resistance.

251. ***The position of resistance is the essential part of the defensive position.***

It is on it that the center of gravity of the defense forces is placed; it is for it that the battle is fought.

Its choice must meet two main conditions:

— to allow, on the zone of terrain where the command has decided to break the enemy attack, the establishment of a barrage of fires from all weapons, known as the *general barrage.*

This barrage must be dense, continuous, and deep. It is formed by the network of combined fires of infantry, artillery, and antitank weapons echeloned over the position to be defended or over the terrain it covers;

— to protect the defense, against the penetrations of the assailant and, in particular, against his armored vehicles, by the presence of a natural or artificial *obstacle* combined with destruction and beaten by the barrage.

In addition, the position of resistance must:

— benefit, in front of the main line of resistance, from open fields of fire, especially in the absence of a natural obstacle;

---

[1] *Ligne d'arrêt.* The U.S. military equivalent of this term is "regimental reserve line" or "final protective line." The British used the term "Brigade reserve position."—Trans.

— cover or, at a minimum, encompass observation posts giving views forward and to the full depth of the position of resistance;

— have concealed communication routes in its rear part.

The *main line of resistance* marks the front limit of the troops in charge of defending the position. It is on it that the main effort of resistance must be directed.

The *stop line* must make it possible to break the momentum of an enemy who would have breached the position, cover the artillery and the rear, and serve, in whole or in part, as a base of departure for counterattacks intended to recapture the lost parts of the main line of resistance.

The heavy weapons that defend it participate, as much as possible, in reinforcing the general barrage.

The main line of resistance and the stop line are connected by *switch trenches* (*cloisonnements*), making it possible to channel the enemy's advance and deny him, in the event of success, from lateral exploitation.

Outposts.

252. The position of resistance is, in general, covered by a system of outposts.

The role of the outposts is:

— to watch the approaches of the enemy and to inform the command;

— to protect the garrisons of the position of resistance against the heavy weapon fires of the enemy infantry and the incursions of armored vehicles;

— to give them the necessary time to establish their combat dispositions;

— possibly, to participate in the mission assigned to the position of resistance.

These missions are, in principle, divided between two echelons: the *surveillance echelon* and the *resistance echelon*, but circumstances may lead to establishing only one echelon responsible for simple surveillance.

The locations and composition of outposts can be different day and night.

The locations of the resistance echelon should encompass observation posts that would be likely to give the enemy close views, offer good fields of fire, allow for the collection of the surveillance echelon, and be themselves supported by fires from the position of resistance, especially by those of some specially designated artillery units.

Detachments or units of artillery can also be deployed in front of the position of resistance and behind the outposts to strike the enemy from as far away as possible.

It is prescribed to strengthen the antitank elements of the outpost system during the initial period of the organization of the position of resistance.

Outposts can, in some cases, participate in the mission of the position of resistance in view:

— either to temporarily retain important points of the terrain in front of the position of resistance (observation posts, strong points, making it possible to temporarily fill in certain gaps in the general barrage, etc.);

— or to break up the momentum of the enemy attack.

The mission of the outposts, their nature, and their strength are variable according to each particular situation of the Large Unit.

It is, in all cases, clearly fixed by the command which is guided, in this respect, by the state of progress of the work of organization of the position of resistance and the deadlines in which the garrisons of this position are able to make their combat arrangements.

253. When the necessary space and means are available, it is advantageous to push beyond the outposts some detachments (cavalry, mechanized or motorized units) equipped with antitank weapons, whose mission is to reconnoiter the most distant enemy, in conjunction with the air force, and to link up with its advance by slowing it down throughout the depth of the battlefield, in particular by disputing as soon as possible with the assistance of the artillery, the points of obligatory passage.

These detachments also provide coverage for the implementation of the destructions.

III. Defense organization and preparation.

254. If the defensive attitude limits the leader's initiative, it at least allows him and even makes it an absolute duty for him to push, as far as the circumstances allow, the defense forecasts.

These forecasts form the subject of the *defense plan* (No. 19).

This plan, established according to the leader's *idea of maneuver,* is based on the choice of directions to be forbidden to the enemy, either because of the advantages that the latter could find there, or because of the possibilities that the leader will reserve for a later resumption of the offensive.

The defense plan defines, in accordance with the general indications developed below, the mission of the Large Units, the defense positions, the general disposition, the conditions of the defense against armored vehicles, the use of artillery and destructions, the missions of air forces and antiaircraft defense, signal communications, and liaisons, measures for the protection of the flanks, as well as the locations of reserves and any forecasts made for their use.

The plan also determines the order of priority of the work and mentions, where applicable, any reinforcements or withdrawals.

It provides for the organization and implementation of services.

Intelligence and Observation.

255. The intelligence is sought as stated in Title III (No. 121). Observation is of considerable importance in the defensive. The plan for collecting intelligence includes, for this purpose, an observation plan whose object is to carry out, by the combined action of all means, a permanent observation, without gaps and as extensive as possible over the entire depth of the position and to organize the safe dissemination and rapid exploitation of the observations made.

General disposition.

256. Large Units are placed abreast in order to promote defense and allow the exercise of command in depth.

Their zones of action are determined by the leader according to their missions, the compartmentalization of the terrain, the respective importance of the directions they deny, and the communication routes. The liaisons between the Large Units are specified.

The protection of the flanks is ensured by the organization of the terrain, by destruction, and by the deployment of the reserves.

Each Large Unit spreads out ***in depth*** in its zone of action, the main body of its forces in the position of resistance, and the reserves arranged behind this position. Inside each Large Unit, the deployment of the troops in the position of resistance is designed in the form of points of force: strongly held *centers of resistance* and *strong points,* well protected by a system of crossfire, surrounded by monitored *curtains,* covered with fire, and barred with obstacles.

The distribution of the occupation of the position of resistance between the different adjoining units results from the prior determination of the points of force to be held.

Each unit supplies the outposts in its zone of action.

257. The *artillery* of each Large Unit is divided into two detachments: one called *direct support,* made available to subordinate units, and the other called *general action,* whose use the command reserves.

This last detachment participates in the defense of the position of resistance subject to being able to promptly execute the fire maneuvers prescribed by the command to ward off the development of the enemy attack.

The general disposition of the artillery is echeloned in depth under the cover of the position of resistance[1] so that the main mass can intervene in the assembly and departure zones of the enemy attacks and that all the matériel can participate in the general barrage and act, in principle, in front of the stop line.

258. *The reserves* are echeloned in such a way that they can intervene quickly to maintain or restore the integrity of the position of resistance.

They include, whenever availability permits, tank units whose use is particularly prescribed in general counterattacks.

Defense against armored vehicles.

259. The defense against armored vehicles is carried out ***over the entire depth*** of the battlefield (No. 420).

The air force, the cavalry, or, if necessary, the detachments referred to in No. 253 and the artillery take part in it.

As soon as the vehicles enter the outposts' zone of action, the artillery and antitank weapons placed at the outposts take them to task.

The fight enters its decisive phase when the vehicles approach the position of resistance. The defensive fire plan provides for this purpose, using antitank weapons and artillery and combining their actions with the ***obstacles.*** Tank units held in reserve stand ready to conduct offensive actions against armored vehicles broken up by previous actions.

The protection of the flanks and the rear is carefully planned because of the extent of the radius of action of current vehicles.

Use of artillery.

260. The missions of the artillery are as follows:

— prevent the adversary from approaching within assault distance of the position of resistance and, at the very least, break up his disposition;

— if, however, the enemy succeeds in attacking the position of resistance, participate with all its fires in the general barrage and ensure, if necessary, the continuity of the fight throughout the depth of the position;

— seek the destruction of armored vehicles in all phases of the battle.

261. The first of these missions requires from the artillery a vigilant and unwavering observation and flexible maneuvers of its matériel and its fires. It can yield decisive results.

---

[1] Except, the detachments temporarily pushed forward from the position of resistance as described in No. 252.

The artillery endeavors to seize, as far as possible, all the vulnerable targets by firing from long-range matériel and by that of mobile matériel deployed temporarily in front of the position of resistance (No. 252).

When the enemy attack seems imminent, the artillery can be called upon to open counter-preparation fire, which aims to seize the enemy at the moment when he takes up his attack disposition.

262. During the attack on the position of resistance, the artillery participates in the general barrage by its stopping fires (*tirs d'arrêt*), executed instantly, as close as possible to the main line of resistance, by all the defense matériel, firing at the maximum rate. They are renewed as appropriate.

The purpose of these stopping fires is to:

— to reinforce the infantry fires or fill gaps in them and thus increase the depth and power of the general barrage;

— to deny, in combination with obstacles or destruction, certain corridors to enemy tanks and reserves.

Inside the position of resistance, the artillery action becomes more difficult due to the absence of precise intelligence, but it can be very effective if one manages to maintain the continuity of the observation.

263. The artillery use plan also includes a counterbattery plan and an interdiction plan.

Organization of the terrain and Destructions.

264. The organization of the terrain is the essential complement of any defensive organization. Its development varies with time and the means at our disposal.

The development of fields of fire, camouflage, the creation of obstacles, antitank in particular, the protection of the sources of fire, and the means of signal communications are undertaken first.

The organization continues within the framework of an overall plan known as the *organization plan* established as soon as possible and which, taking into account the means and the manpower gradually available, provides, in an order of priority, work intended to reinforce the action of fire, protect personnel, facilitate the exercise of command, and feed the battle.

265. The purpose of the destructions is to slow down the enemy's operations by preventing him from using the communication route network (No. 333).

They are particularly effective in terrain with broken ground or poor in communication routes.

Destructions are, as far as possible, combined with obstacles and linked to the disposition of aviation, artillery, and infantry fires.

Carried out beyond the outposts, they delay the enemy's approaches; forward and inside the position of resistance, they reinforce the value of the general and interior barrages.

Only destructions covered by fires affecting extended zones are truly effective and likely to delay the enemy for any appreciable time; their preparation requires significant resources and time.

IV. Command action.

266. The command, after having organized the defense, makes its action felt during the battle, mainly by the maneuver of the fires of its artillery, by the intervention (the aviation (No. 295 and following), and by the engagement of its reserves (infantry and tanks).

Maneuver of fires.

267. During the distant fight, while continuing to execute the artillery fire planned in this phase, the commander of the Large Unit endeavors not to prematurely reveal to the enemy the deployment and strength of his artillery.

It is then up to him to prescribe the opening of the counter-preparation fires. Counter-preparation can be decisively effective if it strikes at the right time and in the right place, but it reveals the strengths of the defending artillery and consumes a large amount of ammunition.

When the enemy enters the position of resistance, the command maneuvers the artillery fire to benefit the particularly threatened parts of the front.

If the position of resistance is broken, he devotes all the available artillery to the mission of supporting and protecting the counterattacks.

Use of reserves.

268. The commander of the Large Unit uses his reserves either to limit the enemy's local successes or to restore the integrity of the position of resistance.

269. In the first case, the reserves are used to reinforce the threatened parts of the position of resistance that give an essential benefit for the defense. He first tries to consolidate the flanks of the pockets created by the assailant in order to prevent lateral exploitation and thus force the attacks to dwindle. He then tries to seal off the base of the pockets and then progressivly restore the continuity of fire by occupying the strongest points located behind the line reached by the enemy.

270. In the second case, the reserves are used to retake parts of the position of resistance conquered by the enemy through limited-strength and limited-range offensive operations, known as *counterattacks.*

The reserve detachments of the front-line units execute *immediate counterattacks* before the *enemy has recovered.*

Large breaches in the position of resistance require larger counterattacks which, generally speaking, are prepared, conducted and exploited as attacks.

Because of the circumstances under which counterattacks are undertaken, and the purpose for which they are intended, simplicity in their conception and vigor in their execution are important factors in their success.

Tanks, by the psychological and matériel effects that their action produces, are particularly prescribed to participate in it.

The deployment of the troops responsible for the counterattack is carried out under cover of a base of departure capable of being occupied in good time.

The artillery, deployed as much as possible in the axis of the counterattack, supports the infantry and tanks under the same conditions as during attacks.

271. In certain favorable circumstances, the command may decide to go on the offensive to benefit from an advantageous situation and the attrition of its adversary.

The general actions undertaken then take the name of *counter-offensives.*

V. Higher command forecasts.

272. Whatever the time and the means that may have been available to bring the organization of a defensive position to a high degree of solidity, the possibility of its breach by the enemy must never be ruled out.

It is up to the ***higher command,***[1] ***and to him alone,*** to guard against such a risk, to provide for the organization of other positions, and, if necessary, to occupy one of them.

This forecast by the higher command should not in any way weaken the firmness of the mission of resistance assigned to the troops in the first position.

Each of the planned positions must be at a distance from the previous position such that the opposing artillery cannot, without moving, neutralize these two positions during the same preparation. Each position must also rely on an obstacle in order to ward off a surprise action resulting from the rapid exploitation of a success obtained on a previous position by an adversary armed with armored vehicles.

---

[1] See Part VI (Army), Nos.319 and 330.

The increased ranges of artillery and the speed of modern vehicles therefore lead to the separation of successive positions by greater distances than in the past.

Two successive positions can be connected by *switch positions (positions en bretelle),* drawn transversely. These switch positions, also advantageously covered by a natural obstacle, make it possible to limit a breach by restoring the continuity of the front. They provide bases of departure for the counterattacks directed on the flanks of an assailant who has broken the first position.

The successive positions and switch positions are occupied at the behest of the higher command by reserve units or by new Large Units.

ARTICLE 2.

WITHDRAWAL FROM ACTION.

273. The withdrawal from action consists of the use of successive echelons installed sheltered by terrain breaks or obstacles in positions as favorable as possible from the actions of distant fires. Each echelon avoids, in principle, close combat and, except in the case of very covered or broken terrain, falls back, preferably during the night, by large bounds, on the next echelon.

The cavalry, the mechanized or motorized units, the artillery, and the aviation are the main elements of the withdrawal from action; their action is advantageously combined with that of destruction.

The cavalry, and even better mechanized units, are used to maintain contact with the enemy, cover the withdrawals and flanks of the echelons, harass the enemy, and even counterattack him during his advance.

The distant fires of the artillery and the powerful actions of the aviation, combined with the obstacles and destruction, hinder the enemy's advance day and night.

274. During the withdrawal from action, the means must not be evenly distributed over the different successive positions, which would lead to a dispersion of efforts detrimental to the proper execution of the mission received.

The density of occupation of each of these positions must, on the other hand, be nuanced according to the duration of the halt that the commander intends to impose on the enemy and be determined according to the time limits allotted to him, the facilities defense offered by the terrain, and the situation of the opposing forces.

ARTICLE 3.

RETREAT.

275. When the battle, offensive or defensive, ends in failure and retreat is necessary, the command endeavors, under cover of rear guards, to withdraw the main body of its forces to a distance from the enemy sufficient to reconstitute units and receive reinforcements.

The rear guards are deployed far enough from the line of battle to set up their fire system before the enemy approaches.

They are composed and operate as indicated in No. 145.

Under their cover, the command organizes without delay the clearing of the rear of the battlefield, in order to free the road network as quickly as possible and then ensure the passage of combat troops without disorder.

Darkness is used to operate the disengagement of the engaged elements. There is no point in sending fresh units in front of withdrawing units to stop their recoiling movement; it is preferable, on the other hand, to establish the fresh units on a position chosen far enough back to allow them to have the time necessary for their solid installation on the ground.

ARTICLE 4.

SIGNAL COMMUNICATIONS IN THE DEFENSIVE BATTLE.

276. *In the defense without thought of retreat,* the organization of signal communications is designed in accordance with the leader's idea of maneuver; it mainly applies the means to the directions that need to be interdicted.

It is also characterized by a reinforcement of means in favor of the position of resistance and, on this position, in favor of the main line of resistance.

It must allow the instantaneous action of the counter-preparation, the general barrage, and the rapid engagement of the reserves.

It must be able to adapt to the various modalities of the reinforcement plan and respond to the most likely eventualities. In particular, in anticipation of penetration of the enemy in the position, the echelonment of the signal communications towards the rear is prepared in such a way as to ensure the continuity of the relations necessary for the exercise of command.

The telephone (or telegraph) network is the backbone of signal communications. As long as the enemy attack has not started, the use of radios is subject to severe restrictions.

The requirements concerning the organization of signal communications are included in the defense plan.

During the withdrawal from action, signal communications are organized at each position according to the mission assigned to the occupying troops.

In the retreat, the signal communications effort is focused on the relations necessary for the rear guards.

Generally, radios, agents of fast signal communications, and the existing telephone network can satisfy the most important needs.

ARTICLE 5.

SPECIAL CASES OF THE DEFENSIVE.

I. Defense of fortified fronts.

277. The organization of the fortified fronts is characterized by the presence of permanent fortification *ouvrages*, which have numerous infantry and artillery weapons and are capable of cooperating powerfully, particularly with flanking fire, in the defense of these fronts.

The position of resistance of the fortified fronts includes the *ouvrages* and the *strong points of the interval troops.* It extends in depth from the main line to the stop line.

The *main line of resistance* has as its backbone *the most important ouvrages of the position;* it is generally covered by a natural or artificial obstacle, if necessary, supplemented and reinforced by antitank mines.

The *ouvrages* occupy the strongest points of the terrain, the possession of which is necessary for the attacker to continue his advance. As long as the *ouvrages* hold, the assailant, even if he succeeds in penetrating the gaps, will only be able to obtain partial and temporary successes.

The *ouvrages* thus become the stake of the struggle; ***on the fortified fronts, it is for the line of ouvrages, the main line of resistance, that the battle is fought.***

The defensive maneuver on the fortified fronts necessarily takes the form of the defense without thought of retreat.

It is designed and conducted according to the general principles of maneuver on a defensive field position. In application, only the characters of the permanent organization intervene.

Defense of ouvrages, intervals and flanks.

278. ***The ouvrages and the interval troops support each other.***

279. *Ouvrages* constitute combat units. They have a combined arms garrison placed under the orders of a leader: the *ouvrage* commander.

The *ouvrage* is normally resupplied from the rear, but it contains important supplies that allow it a long resistance even in the event of encirclement.

Each *ouvrage* must ensure a double mission:

— the fire *support* of the neighboring *ouvrages* and that of the intervals;

— the *retention* of the terrain it occupies.

The first of these missions is specified by the higher command; it defines the fire actions of the *ouvrage* in the overall defense plan. This mission, to which most of the armament is assigned, is ensured *by priority* even if the *ouvrage* is attacked and surrounded.

However, when the life of the *ouvrage* is at stake, the commander, responsible for his *ouvrage*, becomes the sole master of the defense and has full authority over all the means to conduct the fight. Once the crisis is over, it has an imperative duty to return to its primary mission immediately.

The *ouvrage* commander extends the resistance of the *ouvrage* until all means are exhausted.

280. The defense of the *intervals* is organized and conducted *in such a way as to support the ouvrages.* For this purpose:

The interval troops reinforce with their fires the front and the flanks of the *ouvrages.*

Inside the position, fire barrages are planned, and switch trenches are prepared to cover the flanks and communication routes of the *ouvrages*; the layout of the stop line is also guided by the concern to protect, at least by fire, the entrances and communication routes of the *ouvrages.*

281. The *flanks* of the fortified regions are organized defensively.

Their organization makes it possible to protect the fortified region's communication routes and preserve the connection with the neighboring troops, whatever the fluctuations of combat in the regions' intervals.

Use of Artillery and Observation.

282. *Artillery* outside the *ouvrages* receives the same missions as on a defensive campaign position and is articulated similarly.

The artillery of the ouvrages, under casemate or turret, ***has the defense of the main line of resistance as its essential mission:***

It can also be used in distant action; but whenever a mission of this nature can be carried out either by artillery of the *ouvrage* or by artillery external to the *ouvrages*, it is to the latter that preference will be given in order to spare the artillery of the *ouvrages* and the divert as little as possible from its essential mission.

Due to the extent and depth of the zones of action of the artillery of the *ouvrages*, the command always takes care to identify the nature and the *order of priority* of its missions.

For the same reason, the artillery of the *ouvrages* is, in general, advantageously articulated in general-action groupings dependent on the commander of the artillery of the fortified sector. In the frequent cases where these groupings are given the task of reinforcing the direct support of one or more subsectors, the corresponding fires of the *ouvrage* artillery are made available to the commanders of the direct support groups of the subsectors concerned.

283. The fortified fronts have numerous and protected observation posts.

These observation posts are mostly distributed between artillery and infantry, but all work for the benefit of the command.

Defense against armored vehicles.

284. The presence of obstacles and minefields, as well as the variety of the *ouvrages*' armament, facilitate defense against armored vehicles.

It is organized, as in defensive field positions, in the entire depth of the disposition up to the artillery and command posts.

In certain parts of an easily traversed front, a rear barrage can be established on a terrain break or supported by natural obstacles to give the defense even more depth.

Air defense of fortified regions.

285. Air defense is organized, in close liaison with the D.A.T. established in the fortified regions, under the general conditions laid down in Nos. 301 and later.

It is conducted within the framework of the army operations and aims to protect against the aerial dangers:

— the troops stationed in the position outside the *ouvrages*;

— the important components of the *ouvrages*;

— the sensitive points in the rear and especially the supplies for the parks and depots.

Command and Reserves.

286. A Fortified Region is divided into Fortified Sectors.

A Fortified Sector is normally occupied by a Fortified Region Brigade, possibly reinforced.

Each sector is divided into subsectors and neighborhoods.

The *ouvrages* are integrated into this organization. Depending on their importance, they depend on a more or less high level of command.

The command sees to it that the tactical links are and remain solidly established between the *ouvrages* and the interval troops.

287. Fortification allows savings in manpower on the positions but, to conduct the battle, it is essential to have numerous and mobile reserves, Large Units, and elements of the General Reserves. These reserves are articulated in particular for the defense of the flanks of the fortified regions.

Immediate counterattacks.

288. If the enemy penetrates the position, the defense is conducted with a view to clearing, as soon as possible, the overwhelmed *ouvrages* or the invested casemates on which the enemy is concentrating his efforts.

The aim of the immediate counterattacks is then the *complete restoration of the initial front,* which requires more force on fortified fronts than on a defensive field position.

Prepared by the fires of the stop line and by those of the artillery outside the *ouvrages*, these counterattacks are powerfully aided by the flank and reverse actions of the collateral *ouvrages* on the enemy elements that have ventured into the position.

Reinforcement.

289. A *reinforcement plan* provides for the entry into line:

— small field units; or

— Large Field Units.

In the first case, the reinforcements are integrated into the existing disposition. The field units are assigned to the defense of the intervals, with the exclusion of that of the *ouvrages*, and provide or reinforce the local reserves.

In the second case:

— *under threat of attack,* there is a need:

— to reinforce artillery fire and antitank fire as a matter of priority;

— to increase the depth of the disposition towards the rear and possibly towards the front;

— to prepare the intervention of the reserves.

— *during an attack,* the Large Field Units act either to seal a breach or to counterattack.

As a general rule, the fortress troops continue to operate the matériel and dispositions of the permanent installations; the field troops are engaged only in the intervals.

The organization of the command, after reinforcement by large field units, is guided by the concern to ensure an intimate link between the fortress troops and the field troops and to achieve a more complete deployment of the command organization.

For this purpose, the command is divided between the leaders of the large field units and the leaders of the fortress Large Units; depending on the extent of the front to which the reinforcement applies and depending on the importance of the field Large Unit called upon to engage, the commander of the latter exercises command of all or only part of the reinforced sector.

To facilitate the task of the field unit commanders, in particular that of the commanders of Large Units, it will be necessary to add to them, at least temporarily, an officer drawn from the staffs or from the units of the fortress troops.

Withdrawal.

290. In the event that operations take place far from the fortified regions or the command would like to use the fortress troops in open country, the latter's withdrawal would then be prescribed; in this eventuality, they would leave behind certain specialists reinforced by personnel from the old classes.

II. Defensive operations on a stabilized front.

291. The defensive operation may include periods of stabilization during which certain large units will have the sole mission of holding a front in contact with the enemy.

During these periods, the command prescribes all measures with a view to organizing and reinforcing the defense of the front for which it is responsible.

The works are developed first on the position of resistance and then on the other positions according to a long-term plan and by exploiting all the resources of the terrain and the fortification; they are conducted in such a way that they can always lend themselves to good use in the event of an attack.

Thus, a solid front will be gradually created, some parts of which may, depending on the terrain, the time, and the matériel available to the occupants, acquire a value close to that of the permanent fortification.

The rules relating to the preparation and conduct of the defensive battle adapt to this situation.

The defense plan contains more detailed prescriptions. The numerical strength can be reduced as organizations improve; we must strive to leave in place only a minimum of forces on condition that reinforcements are provided for if necessary, the methods of which are studied in the most probable hypotheses.

Camouflage takes on particular importance due to the extended proximity of the two adversaries.

The equipment of the position and the installation of the services are developed in such a way as to facilitate the movements and to ensure the maintenance of all the personnel whose use will have been foreseen.

III. Defensive operations on wide fronts.

292. Defense on wide fronts is a special defensive case.

It is used, in particular, on fronts that are relatively passive or covered by large natural or artificial obstacles. But it can impose itself, whatever the terrain, on the first units engaged in front of superior enemy forces, whether to contain them or to slow them down.

One cannot expect, in any case, from a unit engaged on a large front, a prolonged resistance against powerful attacks

The defense on wide fronts comprises two principal modes of action:

— defense of a single position;

— defense by maneuver.

The choice between these two modes of action depends essentially on the mission, the distance from the enemy, the time to be gained, the extent of the authorized withdrawal, the width of the zone of action assigned to the terrain, and the nature of the means at our disposal.

Whatever this choice, the defensive battle on a front requires:

— a particular concern for intelligence;

— the decentralization of command (possibly pushed as far as the constitution of tactical groupings);

— very secure signal communications;

— the absolute rejection of the uniform distribution of means in line, which would condemn us to being weak everywhere;

— reserves as large and as mobile as possible;

— the maximum use of the properties of the armament, the fortification, and the terrain;

— extensive use of destruction.

293. In the case of the *defense of a single position,* the extension of the front depends on the presence and the value of natural or artificial obstacles existing on all or part of the front, on the time and the means at our disposal to organize the position, and also the time during which it is necessary to hold.

Only serious obstacles can permit extended resistance.

The most advantageous means of gaining time is to delay the moment when the enemy will attack the position of resistance. From this moment on, there can only be any question of containing the attack or limiting its progress: the time gained is subordinated to the vagaries of combat.

The extension of the front must not be obtained by the stretching of the small units or the total suppression of the depth of the disposition, but by the more or less dense occupation of the centers of resistance forming the bastions of a sinuous layout whose lightly occupied gaps are covered by the reciprocal flanking fires of the bastions.

The general barrage's value diminishes, so to delay the enemy's arrival and installation in front of the position of resistance, all the usual processes must be implemented: delay elements based on armored vehicles, destruction combined with the action of distant fires, use of obstacles, organization of minefields, and an antitank fire system.

Matériel obstacles and camouflage are all the more useful when the manpower is limited in relation to the front to be held and does not allow the creation of a system of extensive outposts on the entire front.

294. Defense by maneuver is essential as soon as it becomes impossible for the command, due to the extent of the front, to install a disposition of continuous fire.

It consists in holding the front with the minimum of forces necessary to stop the incursions of the enemy and to slow him down, while the main body of the judiciously articulated forces is kept available to counter and possibly respond to enemy attacks.

The maneuver takes two main forms which are not mutually exclusive:

— withdrawal from action;

— maneuver by concentration of forces on the threatened portion of the front. The withdrawal from action is executed according to the ordinary procedures: however, the extent of the front leads, in this particular case, to a large decentralization of command.

The maneuver by concentration of forces takes the following general form:

— in front of each penetration corridor, a combined arms detachment has the task of allowing the maneuver of the main body;

— fixed or mobile surveillance elements link the detachments to each other;

— the detachments withdraw from action on axes fixed by the command;

— in the event of a generalized attack, the maneuver can only be transformed into a withdrawal from action;

— in the event of a localized attack, the main body intervenes to stop the enemy in a reconnoitered position or to counterattack him.

Intelligence, surprise, and speed are the conditions for the success of such a maneuver, which also requires the leader's quick eye, decision-making, and audacity.

CHAPTER IV.

AIR FORCES AND AIR DEFENSE IN BATTLE.

I. Aviation.

295. Like the other arms, aviation makes its capital effort in battle, whether it is a question of informing the command and the troops, blinding the opposing air force, or supplementing the action of the artillery with its own fire as far as the most distant rear areas of the battlefield.

Exceptionally, it may be called upon to land detachments responsible for carrying out certain specific missions on the opposing battlefield.

296. To inform the command, aviation endeavors to seek mainly intelligence that makes it possible to conduct the action and engage the means wisely in the battle. For this purpose, it will apply its research, in particular, to the surveillance of enemy reserves, whose distribution and movements reflect the possibilities of the adversary.

297. As far as the troops are concerned, the air force first informs the artillery: searches for targets and setting up fire.

Given the increasing range of weapons and the increasingly effective measures that the enemy will take to shield his troops from sight on the ground and to reveal only at the last moment the whole of his disposition, the distant reconnaissance of the objectives and the observation of artillery fire can only be ensured in good conditions with the active and intense assistance of the air force. Without this help, the artillery would remain partially blind.

Artillery missions thus constitute the most important task of observation aviation in the battle, working for the benefit of the troops.

On the parts of the front where the ground observation posts are not sufficient, the air force is, moreover, called upon to inform the command on the situation of its first elements and to inform the infantry troops by accompanying them in combat.

298. During certain phases of the battle it is of paramount importance to ensure freedom of action for observation aviation and to shield friendly troops from the sights and attacks of enemy aviation. To achieve this goal, it is necessary to have the mastery of the air.

It is from the combined action of defensive light aviation and heavy defense aviation, the first attacking the enemy in flight, the second on his ground, that this mastery of the air will be required. Moreover, it can only be obtained on fronts and for limited periods of time. The command must consequently decide on the points and times at which this mastery of the air, which then demands the most intense effort from the air force, will be sought.

299. Finally, *by its fires,* the range, suddenness, and power of which produce impressive matériel and psychological effects, aviation cooperates in a formidable manner in the destruction of the enemy to the most distant rear areas of the battlefield.

There too, the command will have to choose the objectives and the points on which the aviation attacks will be concentrated in order to obtain impressive results at these vital points and at the opportune times. It is especially the assembling of troops, the columns marching towards the battlefield and, when the enemy is routed, the retreating troops which constitute the objectives to be fixed to the air force in the zone where decisive effects are sought.

In order to achieve the indispensable rapidity of intervention, it is advisable to leave the initiative for launching attacks to the air force commanders, after having first oriented these commanders on the nature of the objectives to be attacked and the goals to be achieved.

It is also advisable to seek effects as massive as possible by involving in this action by fire, not only the bomber squadrons, but all the squadrons available.

Thus, living with passion the unfolding of the battle, animated by the ardent desire to bring its intelligence and its fires a powerful help to the troops on the ground, the aviation deploys in the battle all its power *by seeking the convergence of efforts* which, alone, makes it possible to obtain success.

300. When the circumstances are favorable, and the importance of the goal in view justifies such a decision, the command may be led to ask the aviation to land, behind the enemy's Large Units, combined arms detachments charged with a specific mission (destruction or occupation of an obligatory crossing point, a communication node, an important supply center, etc.).

Such actions require careful preparation and trained troops for these special missions.

They generally include the prior deployment of paratrooper elements responsible for covering the subsequent landing of the main body of the detachment.

II. Antiaircraft defense (D.C.A.).

301. Antiaircraft defense participates in the battle by providing, through intelligence and fire, in conjunction with aviation, *air cover,* which aims to prevent investigations and attacks by enemy air forces on the battlefield. and on his rear (No. 153).

Antiaircraft action is characterized by its permanence, while that of aviation is intermittent.

It follows that to ensure air cover without gaps, the combination of these two weapons in space must be obtained by superposition of their areas of action rather than juxtaposition.

302. The command sets, depending on the maneuver on the ground, the conditions under which the air cover will have to be ensured during the battle, in both time and space.

To this end, it organizes the combination of aviation and D.C.A. actions (including territorial antiaircraft defense).

The action of the D.C.A., whose concentration of fires is the condition of efficiency, is exerted:

— in the zone of the large first-echelon units by creating a barrage as close to the front as possible and as continuous and deep as possible (role of the *forward groupings*);

— in the rear area, by ensuring the defense of sensitive areas or points (role of the *rear groupings*).

Signaling missions for the benefit of defensive light aviation can be entrusted to the deployed antiaircraft batteries, provided that such missions are superimposed on the cover missions proper and do not lead, on the other hand, to the dissociation of the batteries charged to ensure them.

The air cover of Large Units on the march cannot be ensured continuously; it is limited to defending with the D.C.A. the most sensitive points or zones by devoting all the means there if necessary: the defensive light aviation, to the extent of its availabilities, supervises and protects the columns in march.

This rule applies, *a fortiori,* to motorized Large Units because of the range of their movements and the speed of their advance.

303. Command may:

— reserve the use of forward groupings. They are then attached to the Large Units for their movement, their protection, and their supply; or

— put these groupings under the direct orders of the commanders of Large Units (case of the approach march, exploitation, etc.).

The command always reserves, in principle, the re-use of the rear groupings.

III. Territorial antiaircraft defense (D.A.T.).

304. When the battle is fought on the national territory, all the D.A.T. means (No. 57) established in the Combat Zone take part in it:

— the general security bodies, by alerting then informing the aviation and the D.C.A. of the armies as well as the Large Units;

— units of artillery, machine guns, barrage balloons, and searchlights contributing to the air cover of the armies.

This participation in the battle must not, however, lead to any disruption in the accomplishment of the normal mission of these different means, namely:

— for the general security units (lookout, intelligence, signal communications), collect and transmit, at any time, the intelligence ***on which the antiaircraft defense of the interior zone depends;***

— for the other means, defend as a priority the sensitive points or areas to which they provide protection.

It follows from these reservations that the commanders of Large Units, in particular army commanders, have the obligation:

— not to make any modification to the disposition of the D.A.T. means installed in their zone of action, without prior authorization from the Commander-in-Chief;

— plan and ensure continuity of operation of the general security system in the event of advance or falling bank.

If the advance takes place beyond the national territory, it is up to them to organize new lookout lines as soon as possible, extending the general security system of the national territory into the occupied territory.

If, conversely, the development of the battle involves a rupture of this system, the command must endeavor to restore as soon as possible the continuity of the intelligence, either by using switch position lookout lines, initially envisaged, or by improvising new lookout lines

TITLE VI.

THE ARMY IN BATTLE.

---

305. The army, the *fundamental unit of the strategic maneuver,* is equipped with all the means necessary to continue, *in the tactical order*, the realization of this maneuver.

The Commanding General of the Army, acting within the framework of the mission received from the high command, must, therefore, both design his own operation and ensure its execution. This dual prerogative gives him, better than at any other level, the power to imprint on the action the unity essential to success.

To this end, he gives orders to the commanders of Army Corps and other Large Units directly under his command, to the commanders of the artillery, engineers, air forces, antiaircraft defense, and tanks of the army. He coordinates their action.

306. The conduct of the army in the battle, offensive or defensive, is guided by the general rules formulated in the preceding Titles, in particular in Titles III and V and the indications of a particular order which are the subject of the following chapters.

CHAPTER ONE.

THE ARMY IN THE OFFENSIVE.

ARTICLE 1.

PRELIMINARY ARRANGEMENTS.

307. Immersed in the instructions received and the role assigned to the army in all operations, as well as intelligence on the enemy gathered until then, the army commander designs, in principle, his maneuver *until its conclusion.*

He therefore decides on the plan of maneuver (No. 19), which will serve as a guide for the conduct of operations.

He establishes, for this purpose, the intelligence plan (No. 23) which will allow him to ensure, by successive alterations, the evolution of the maneuver towards the goal he has set himself.

Reconnaissance and Security.

308. The air force and the cavalry jointly ensure *reconnaissance* and *security* under the conditions provided for in Title III (No. 130 and 139).

309. The possibilities of the army-level cavalry, even with the assistance of the reconnaissance battalions of the subordinate Large Units, will not always ensure, concurrently and over the entire the army front, ground reconnaissance and distant security.

In such a case, rather than risking a harmful dispersal of the main body of the army cavalry by assigning it a mission of security over the entire front, it will be preferable, after having allocated sufficient means to the distant reconnaissance, to maintain this main body on the most advantageous axis, often on the direction of the army or of a wing. He will thus have his means sufficiently combined to fight effectively, if his mission requires it, either to support the distant reconnaissance, or to prevent the enemy from crossing a terrain break or to gain time by a withdrawal from action.

Directions.

310. The army commander has no interest in identifying, initially, to the subordinate Large Units directions so extended that they trace his maneuver to the conclusion he has envisaged. Unforeseeable events may, in fact, require modifications to the original plan. They will often result in certain inflections imprinted momentarily on the leadership of one or more army corps. This constant adaptation of the initial arrangements to the realities of the moment constitutes, in the hands of the army commander, a simple and powerful means of making his action felt. This is how a weak convergence imposed on the directions of neighboring army corps very quickly leads to a considerable increase in resources on a given front. Conversely, a slight divergence of directions can make it possible to quickly envelop a region that is difficult to tackle head-on, to decongest an overloaded network, etc.

However, the directions of the army corps must be staked out far enough and extended in good time so that these Large Units are never indecisive.

The directions assigned to the army corps must always, in the end, have the effect of driving the army in its direction.

ARTICLE 2.

APPROACH MARCHES AND DISPOSITION.

311. The approach marches of the army are carried out according to the general rules prescribed in No. 209.

In a supervised army or one whose theater of operations is quite narrowly limited, the first care of the command is to take possession of the whole of its zone of action by having in the first echelon the number of Large Units strictly necessary to constitute, as soon as the need arises, feel *a solid deployment front.* When the army is not yet assembled, the first Large Units available form the backbone of this first echelon of forces and are given the task of covering the army assembly.

In the second echelon are articulated the Large Units and General Reserves whose intervention must ensure the development of the maneuver. It is beneficial for these second-echelon forces to be numerous, because they are the ones who ultimately allow the army commander to impose his will on the adversary.

Far from the enemy, simply supervised intervals can be set up, between the first-echelon Large Units, as well as with the neighboring armies. The disposition gains in flexibility. But all arrangements are planned to restore the continuity of the front, when the contacts become clearer. In this respect, mention is made either of forces originally maintained in the second echelon, or of the main body of cavalry which, in this phase, may no longer find their use in advance of the front.

The second-echelon Large Units are strongly articulated in width and depth, but always according to the role assigned to them in the mass maneuver.

When the army approaches the enemy, the disposition tightens, especially in depth; the place of fast-moving Large Units whose arrival is often expected at the last moment is spared, if necessary.

In all cases, the army commander ensures air cover of the approach marches of the army or addresses, for this purpose, the necessary requests.

ARTICLE 3.

ESTABLISHMENT OF CONTACT AND ENGAGEMENT.

312. The army commander's instructions concerning *establishing contact* define the missions of the distant security units and the first-echelon Large Units.

The advance guards of the latter can be called upon, depending on the attitude desired by the army commander, either to support the cavalry in contact as soon as possible and to reinforce its action, or on the other hand to collect it in a given position.

The first of these missions to be assigned to the advance guards corresponds above all to the case of an enemy in position or withdrawing.

The second is prescribed in front of an enemy on the offensive march, which it is then important to stop before attacking (No. 201 and 219). To this end, the first-echelon Large Units deploy without delay to a position favorable to the rapid organization of an effective fire system and settle there with all the advantages inherent in ***the priority of the deployment.***

This position depends on the battlefield chosen by the army commander and as such can be prescribed imperatively by him; in other circumstances, if the terrain lends itself to it, he leaves the determination to the commanders of the army corps, within the limits he is careful to set for them.

During the operations of establishing contact, the army artillery intervenes gradually.

313. As soon as contacts become clearer, the corps commanders take charge of the conduct of the fight in their respective zones. Within the framework of the orders received from the army commander, each of them proceeds to the *engagement* of his army corps, as indicated in No. 221.

The Commanding General of the Army takes care, from the start of this phase, to orient the army corps engaged on the points which must be seized without delay to facilitate the subsequent development of the maneuver that he is pondering; he sees to it that, even outside the zone where the army attack will develop, the first-echelon army corps not only strongly constitute their line of battle by reinforcing the advance guards, but engage vigorously in well-prepared and well-supported local actions, which are often combined. He provides early support of the army artillery.

If these engagement actions are not enough to break the opposing resistance, they define it at least with precision and, by their distribution even on the whole of the front, maintain the enemy in indecision.

ARTICLE 4.

ATTACKS.

314. During the establishment of contact and the engagement, the army commander carefully examines the results achieved, the intelligence gathered, and the adverse reactions. He confronts these positive data with his possibilities and his intentions; he thus matures his maneuver and then specifies the conditions under which, personally taking the action in hand, he will attempt to break enemy resistance by *attacking,* the *capital* act of battle.

To this end, he determines the part of the front and the axis on which he will make his effort and proceeds methodically to a *distribution of his forces* which allows him to always have, on the different parts of the battlefield, sufficient means to ensure, in spite of the enemy, the success of his attack.

315. The army commander achieves this distribution of forces by a fair *adaptation of the means to the missions* and by a judicious *deployment of the reserves.*

For this purpose, he reinforces certain army corps. He brings into line army corps, divisions and General Reserve units hitherto kept in the rear; if necessary, he refers the artillery of the Large Units kept at the beginning in the second echelon.

He gives the commanding general of the artillery of the army his instructions concerning the distribution of the artillery units of General Reserve between the army corps and the army artillery; he decides early on the role of the latter.

He proceeds, under similar conditions, to the distribution of the tanks placed at the disposal of the army. He prescribes the conditions of their use.

He sets the locations of the army reserves, advantageously chosen from among the Large Units or the motorized or mechanized General Reserve units.

He makes levies from the passive parts of the front and prepares those he will take, as soon as the attack is launched, to push the economy of forces to its extreme limits.

316. The army commander determines the *organization of the command* and that of the liaisons and adapts to the needs of the attack *the search for intelligence* and the methods of its dissemination.

At the same time, he sees to the development of his rear, ensures the establishment of his services and deploys the ammunition and the necessary matériel. He has prepared, by using a traffic plan, for supplying the battle and evacuations.

All these measures are intended to ensure, as quickly as possible, the deployment of *all the means* at the army's disposal and their application in the direction in which success makes it possible to expect decisive results.

317. The attack develops under the general conditions set out in Title V. It is led from start to finish by the army commander, as stated in No. 239.

ARTICLE 5.

COMPLETION OF THE BATTLE.

318. The army commander, according to the results obtained by the attack, organizes the exploitation of the success or, in case of failure, puts himself in a position, after having ensured the solid possession of the conquered ground, to be able to resume the offensive as soon as possible (Nos. 241 and following).

CHAPTER II.

THE ARMY IN THE DEFENSIVE.

ARTICLE 1.

DEFENSIVE WITHOUT THOUGHT OF RETREAT.

319. The role of the army in this form of defensive battle differs essentially from that assigned to the army corps and the division.

These Large Units have no other mission than to *hold,* without thought of retreat, the *unique* position of resistance entrusted to the value of their defense.

The army can and must give the defensive battle a scale commensurate with the importance of its resources.

The army commander achieves this by increasing the *depth* and *flexibility* of the defense, both by using actions carried out *in front of the position of resistance* when the distance from the enemy allows it, and by the development of successive positions organized *behind the position of resistance,* in anticipation of its rupture (No. 272).

This results, for the army commander, in two different orders of concern:

— to prepare and conduct the defense of the *position of resistance;*

— to plan and, if necessary, conduct *the defensive maneuver of the army,* forward and, if necessary, behind the position of resistance.

320. From this double point of view, the army commander establishes the *defense plan* of the army (No. 254) and the corresponding *intelligence plan* (No. 23).

I. Defense of the position of resistance.

Defense organization and preparation.

321. The mission of the Large Units charged with defending the position of resistance of the army is always a mission of resistance without thought of retreat.

322. Unless it is a matter of defending a fortified front, the army commander chooses the position of resistance on the basis of the mission received, the main directions which must be denied to the enemy and essential points which must be preserved.

He determines, in broad strokes, the general outline of the position of resistance, as well as the depth of this position.

This route makes the best use of the advantages of the terrain, in particular the large obstacles and the large uneven ground.

The army commander also defines the general conditions for the establishment of the outpost system and the mission of these outposts.

323. Each army corps, placed on the position of resistance, receives a zone of action corresponding, as far as possible, to a compartment of terrain and to a direction to be interdicted, it is thus able to devote itself entirely to a mission whose *unity* is thus ensured.

The army artillery, largely echeloned, is deployed in such a way as to be able to act deeply and to reinforce the action of the corps artillery.

324. Army reserves are made up of Large Units, motorized if possible, and General Reserve units, including tanks and antitank vehicles.

The reconnaissances of the Large Units and units held in reserve are oriented according to their possible roles. The tasks incumbent on their troops are specified.

In a wing army or placed on the wing of a fortified or stabilized front, a judicious echelonment of the reserves guarantees the uncovered or unfortified flank.

325. If the front stabilizes, the army commander has the duty, as the defensive organization improves, to reduce more and more the number of troops in the sector in favor of the army reserves.

These, made up of General Reserve units and, as far as possible, of complete Large Units (divisions and army corps) will normally be held ready to reinforce the army front in the event of a powerful enemy attack, possibly to be directed on active fronts at the request of the high command, or to take the offensive on the front even the army, according to the directives given by the high command.

The army commander consequently draws up a *reinforcement plan,* possibly a *withdrawal plan* and an *offensive action plan.*

The army front then receives equipment responding to these eventualities and making it possible to absorb, without attracting the attention of the enemy, the considerable reinforcements required by the offensive and to implement them in the minimum of time.

Conduct of the defense.

326. The army commander makes his action felt by the maneuver of fire and the use of his reserves, as stated in Title V (No. 266).

When the attack seems imminent to him, he orders the *counter-preparation*, unless he has previously left its unleashing to the initiative of his subordinates.

He also orders, in principle, the execution of the *destructions* according to the methods provided for in the destruction plan (No. 265).

327. The Large Units carry out the fight with all their means on the position of resistance; the army commander closely follows the combat. He grants, in good time, the necessary reinforcements as well as the reinforcement of the army artillery and air forces, in particular for the benefit of the major counterattacks organized and launched by the commanders of the army corps.

In the case of a rupture affecting the front of several army corps, it is up to the army commander to restore the situation. To this end, he coordinates the fire and the actions of the Large Units which are around the breach and he regulates the use of his reserves himself. The engaged troops cling to the terrain and seek to improvise a continuous line of fire. The army commander, as soon as the situation seems favorable to him, counterattacks the attacking forces (No. 270).

If, in particular, the rupture occurred under the pressure of enemy attacks equipped with numerous armored vehicles, he endeavors to stop their progress, preferably at the height of a terrain break, by using the units on site and fast-moving Large Units. He reinforces them as much as possible with antitank weapons, in order to put them in a position to quickly constitute an antitank barrage. Subsequent counterattacks are advantageously equipped with tanks capable of attacking enemy armored vehicles.

When the situation allows it, the army commander can decide to switch to the counter-offensive.

328. In a protracted defensive battle, the army commander sets the general conditions of the reliefs required to repair the physical and psychological forces of the units; their proper execution requires careful planning.

II. Defensive maneuver of the army.

In front of the position of resistance.

329. Actions carried out in front of the position of resistance, when the necessary space and means are available, are intended to inform the army commander and, if the situation requires it, to cover the defensive position for the time necessary to put it in a state of defense, to slow down the enemy's advance, or even to break up its approach march disposition.

Organized and coordinated by the army commander, these actions utilize the Large Cavalry Units assigned to the army, failing this, the detachments referred to in No. 253.

Back from the position of resistance.

330. If at the subordinate levels of command, it should only be a question of the *defense without any thought of retreating* from the position of resistance, the army commander has the duty to consider the possibility of a rupture of this position, even if it is fortified, and to deal with it by the use of *the successive positions* and the switch positions provided for in No. 272, entitled: *Forecasts of the higher command.*

331. The successive positions and the ramp positions, determined by him, are, depending on the situation, partially or completely organized, simply held by security garrisons or permanently occupied by detachments of Large Units.

In principle, this work and this occupation are entrusted to units completely separate from the army corps in charge of defending the position of resistance.

The artillery of large units in reserve may be temporarily installed in successive positions.

332. The successive positions give the army commander possibilities of maneuver which it is up to him to use if necessary.

It is the comparative examination of the mission received, of the intelligence gathered on the possibilities of the enemy, of the means at his disposal and, if necessary, of the evolution of the situation, that indicates to him the moment when it becomes opportune to maneuver on the successive positions and on the switch positions.

He uses these positions, either to bring back his main resistance to one of them, or to withdrawal, or to cover the retreat.

But these maneuvers must, in order to bear fruit, have been foreseen and prepared in all their details. Last-minute modifications to a defensive disposition are extremely dangerous.

ARTICLE 2.

WITHDRAWAL FROM ACTION.

333. The withdrawal from action is carried out under the general conditions provided for in No. 273.

It is combined with rendering the communication routes abandoned to the enemy unusable (No. 265).

To this end, the army commander adopts the *destruction plan,* on the proposal of the army engineering commander, and sets the conditions for its implementation. In principle, it is up to the army commander to order the execution of the destruction, but he can delegate this right to the commanders of the Large Units under his orders, identifying the conditions under which they will have to use this delegation.

ARTICLE 3.

RETREAT.

334. The army commander is guided in this respect by the rules set out in Nos. 145 and 275.

His main concerns are to maintain or restore order, to cover the flanks of the army and to liaise with neighboring units.

The army commander indicates, for this purpose, the directions and the zones of action of the army corps, distributes the routes, prescribes all useful measures to lighten the columns and supply them, ensures that the movements of the aviation are carried out under conditions which make it possible to ensure the permanence of its intervention.

He sets a line on which the army corps will establish rear guards intended to allow the passage of their main bodies. He demands from the troops of all the arms that constitute the rear guards (cavalry and tanks in particular), as well as from all available air forces, the sacrifices that the situation entails.

TITLE VII.

THE ARMY CORPS IN BATTLE.

---

335. The Army Corps is *the battle unit,* capable of carrying out, within the framework of an Army, extended tactical action and leading it to a decision.

It groups together a variable number of divisions and organic or reinforcements of all kinds (arms or services).

Thus, the Army Corps is a tactical executing body for the army commander and, for the units he supervises, a *command body* responsible for directing and coordinating their actions.

336. The Army Corps commander is initiated into the army commander's plans by a personal instruction which informs him of the goal, the form, and the rhythm of the maneuver that the latter undertakes and which defines his mission in this maneuver. The mission of the Army Corps is always indicated on the ground by:

— a general direction to follow or to interdict;

— a zone of action in width or in depth corresponding to this direction;

— objectives to conquer or positions to defend.

The Army Corps commander is, moreover, driven by orders that determine step-by-step the main lines of his operations by harmonizing them with those of the neighboring Large Units.

337. Duly informed of the intentions of his commander and imbued with the meaning of his mission, the Army Corps commander designs and decides on his own maneuver under the conditions set out in No. 7.

His decisions mainly relate to:

— the conditions for the deployment of the main body of the Army Corps;

— the dispositions suitable for this implementation;

— the missions of the immediately subordinate elements: air forces, cavalry, divisions, corps artillery, and engineers, etc.;

— the materialization of these missions on the terrain: directions, zones of action, objectives, or defensive positions.

338. The task of the Army Corps commander is both coordination and direct action.

He coordinates the operations of his divisions, in order to obtain joint actions and to achieve, at essential points and moments, the concentrations of effort and, through them, the *mass effects* that lead to success.

He acts with his own organic elements, either to inform, cover, or prepare; or to reinforce, support, or extend the actions of his divisions.

CHAPTER ONE.

THE ARMY CORPS IN THE OFFENSIVE.

ARTICLE 1.

APPROACH MARCH.

339. The Army Corps carries out its approach march according to the general rules prescribed in Title V (No. 209 and following) and within the framework fixed by the army (No. 311).

The approach march disposition presents in the first echelon the number of divisions necessary to constitute, as soon as contact is established, over the entire width of the zone of action a solid deployment front. The forces maintained in the second echelon vary with the situation: an Army Corps called upon to provide an extended effort will have to preserve significant reserves, which could include entire divisions; a wing Army Corps will most often march in successive divisions, more or less echeloned, so as to support its outer flank and remain in a position to easily change direction.

The service units of the first-echelon divisions can be pushed behind the combat units of the second-echelon divisions in order to make the entry into action of the latter more rapid and easier. For the same reason, the second-echelon divisions will benefit from pushing their artillery to the head of their columns.

During the progression, the reorganization of the disposition could be controlled by the evolution of the tactical situation: the Army Corps commander will proceed by modifying the zones of marches, the echelonment of the divisions, or the deployment of his organic elements.

340. To move his disposition, *by day* or *by night,* with order and precision, the Army Corps commander sets the successive lines that the heads and tails of the main bodies must reach, as well as the resulting objectives for the advance guards.

341. *At night,* the main bodies move under the protection of distant security elements, provided by the army or the Army Corps (reconnaissance battalion, reinforced if necessary).

These security dispositions are set up before nightfall on cover positions using natural obstacles.

The columns are distributed in the zone of movement so as to reduce the duration of their total passage, thanks to judicious use of the road network; all forecasts are made in view of the route variations that the effects of enemy bombardments might impose.

The necessary arrangements are also made to hide the elements of the columns whose movement ends during the day from enemy observation.

342. *During the day,* the use of paths concealed from the sight of the adversary is essential and progressively leads the troops into the compartments of terrain favorable to their entry into action.

The Army Corps marches informed by air forces and cavalry, protected by defensive light aviation combining its action with the means of the D.A.T. and D.C.A., covered from *afar* by distant security elements, *closely* by its advance guards.

Corps aviation operates up to the short limit of the zone explored by army aviation. It conducts its investigations methodically according to a research plan notified and kept up to date by the Army Corps staff. It must be informed of the conditions under which the army aviation acts in front of it, of the protection it can expect from it as well as of the conditions for the implementation of antiaircraft defense artillery.

The Army Corps reconnaissance battalion guarantees the distant security of the disposition if the army has not provided for it, and as long as the distance from the enemy makes it necessary. It achieves this:

— by concentrating the effort of its research in one or more well-defined directions.

— by occupying, until the arrival of the advance guards (or flank guards), such terrain irregularities whose possession will cover the advance of the Army Corps.

Its means of command also allow it to be reinforced by detachments of motorized infantry and artillery. In any case, he operated in close liaison with the air force and with the army cavalry when the latter preceded it.

The reconnaissance battalions from the first-echelon divisions remain, in principle, at the disposal of these Large Units. However, there are circumstances that justify the temporary assembly, under the same command, of one or more divisional reconnaissance battalions and the Army Corps reconnaissance battalion: for example, to reinforce and coordinate the security network in front of the first-echelon division which follows the main direction of the Army Corps—or to increase the cover on an open flank of the establishment.

343. The Army Corps commander sets the general mission of the advance guards and specifies the conduct to be followed in the presence of the enemy and the conditions of their entry on the battlefield that the army envisages. Consequently, he also determines, with some exceptions, their composition and strength, which must correspond to the nature and extent of the planned efforts.

The bounds of the advance guards correspond to the successive objectives fixed for the advance guards by the Army Corps commander (No. 340). Intermediate bounds may also be prescribed on the initiative of division commanders.

344. The organic elements of the Army Corps and, if necessary, their reinforcements occupy the position in the system that best corresponds to the timeliness and effectiveness of their intervention against opposing aviation, artillery, and armored vehicles.

Antiaircraft artillery is deployed to provide air cover, with the help of aviation and D.A.T. (No. 153).

Army Corps artillery units are temporarily detached near the first-echelon divisions for counterbattery action against enemy artillery as soon as possible, with the assistance of aviation.

Tanks suitable for combating armored vehicles may be distributed among first-echelon divisions or held in the Army Corps reserve.

ARTICLE 2.

ESTABLISHMENT OF CONTACT.

345. *Contact* is established first by Army Corps cavalry (Army Corps and divisional reconnaissance battalions), more or less sporadically, and then by advance guards, which, under the direct impetus of division commanders, completes and gradually tightens it on the entire march front.

Its importance is paramount because it provides the command with concrete data, without which it risks striking its first blows in a vacuum or in error.

346. *In front of an enemy in position,* the Army Corps cavalry endeavors to drive back the advanced elements of the enemy that it meets in the vicinity of the axes of advance that were assigned to it. When it is immobilized, it keeps in touch, warns the advance guards and observes, ready for action. The advance guards join it and, putting all their means to work, attacking and infiltrating, seek to push forward in order to determine the enemy's first line of resistance, usually evidenced by a continuous network of fires. This result obtained, they establish themselves on the terrain to cover the deployment of the main bodies.

347. *In front of an enemy on an offensive march,* the primary goal is to stop him before attacking him (No. 312).

Directed by the army commander on the battlefield where he is planning the encounter and on the position that the first-echelon Large Units will have to use for their preliminary stopping action, the Army Corps commander sets the successive lines of the terrain which will mark the advance of his divisions until they have reached the position prescribed.

He sees to it that the advance guards of the first-echelon divisions have a composition, a strength, and an articulation that enable them to take on the enemy, from the outset, ***the priority of the deployment and of the adjusted fires*** throughout the width of the zone of action of the Army Corps.

He sets the nature of the cooperation of the advance guards and cavalry depending on the mechanism of the march and the characteristics of the terrain. He was responsible for ensuring that the advance guards are effectively led by their division commanders, and he ensures that the latter arranged their main body in the order most favorable to the active and sustained development of the battle.

He himself follows, very closely, the play of the advance guards of which he takes the high direction as soon as they deploy for their stopping action.

ARTICLE 3.

ENGAGEMENT.

348. The Army Corps commander carefully monitors the operations of establishing contact so that, in accordance with the army instructions, the *engagement* can proceed without delay. (No. 313)

The engagement is the work of a first-echelon forces of variable composition, which generally includes, in addition to the advance guards, the maximum amount of artillery, and a detachment of the main body's infantry, supported, if possible, by tanks.

The Army Corps commander directs the engagement, according to the orders received, the intelligence gathered, and the terrain; he undertakes it on the whole of the zone of action or only inside a favorable compartment of this zone. In the latter case, he gives the elements that are not in charge of it the necessary orders so that they support the engaged troops with their fires and profit without delay from their successes to advance in parallel. He sets for the execution divisions their axis of effort, their zone of engagement, the objectives to be conquered, and the action to be taken after this conquest. He supports or reinforces them with the artillery at his disposal. He sees to their reciprocal support if necessary.

His aviation, while pursuing the search for intelligence, in particular on the enemy defense system (photographic reconnaissance), mainly provides observation missions for the benefit of the artillery and support missions for the benefit of the divisions. His aerostation comes into play to the necessary extent.

At the same time, with a view to the subsequent attack by the main bodies, he tightened up his disposition, modifying them as needed, and had his supply services activated, beginning with that of ammunition. In principle, he will not proceed to the execution of this attack until he has gathered sufficient means and on the order or the consent of the army commander.

The engagement, whatever its purpose, must ensure that the Army Crops deploys a coherent front across the entire width of the zone of action and, consequently, the possibility of maneuvering its main forces under cover.

ARTICLE 4.

ATTACKS.

349. According to the results obtained by the engagement and in accordance with the orders received, the Army Corps commander establishes the plan of attack.

This plan organizes, in the most successful direction, a main attack with maximum forces. The remaining means are assigned to actions that support or cover this attack.

350. To break the enemy disposition, the attack must penetrate at least as far as the artillery positions and be capable of developing without delay. Its direction and purpose are set accordingly.

The Army Corps commander determines its modalities (No. 352) after having precisely assessed the effects to be expected from its means of fire and armored vehicles.

To this end, it seeks, mainly through air forces, as complete intelligence as possible on the armament and organization of the enemy troops as well as on the ground they occupy.

351. When the engagement has given positive results, the attack will advantageously follow it as soon as possible in order to hinder the enemy's freedom of action.

A delay, however, will often be necessary, either to complete, by aerial and ground observation, the study of the enemy's defensive system, either to incorporate the elements of reinforcement granted by the Army, or finally to proceed with the artillery destructions to open the way for the infantry and tanks.

These preparatory operations should then be accelerated to reduce the time available for the enemy to organize his own parade.

352. The orders given by the Army Corps commander relate essentially to the points listed in No. 337 and, in addition, to the following:

— conditions of preparation and execution of the debouchment of the attack;

— determination of the successive objectives on which the entire system will be taken over or reorganized for a new phase of efforts and which, as a result, are said to be *Army Corps objectives;*

— reciprocal support of first-echelon divisions;

— possible operating directions;

— role of elements initially reserved.

The Army Corps commander attaches the greatest importance to the search for intelligence, a vital element of decision-making. He intervenes personally for this purpose to identify to the air forces their mission of research during the various phases of the attack.

Because of its importance and its complexity, the *emplacement* of the attack disposition on the base of departure, facing its first objective, will generally be the subject of a special order detailing all the measures required to guarantee its secrecy, methodical execution, and protection.

The emplacement can only be carried out by day if the Army's defensive light aviation and its antiaircraft defense forces provide effective cover.

353. The Army Corps commander takes responsibility in principle, with the artillery at his disposal, for counterbattery and distant interdiction; he sets to this artillery its missions and its distribution according to the proposals of the Army Corps artillery commander. The latter consequently constitutes the groupings and places them on the terrain under the most favorable conditions to prolong and, if necessary, reinforce the action of the divisional artillery.

354. When tank formations have been assigned to the Army Corps, their use follows the rules set out in No. 230.

The Army Corps commander, after having reserved certain tank units if he deems it useful, distributes the tanks of all models between the divisions, taking into account the latter's missions and the nature of the terrain of their zone of action.

He specifies when the accompanying tanks are to be deployed and, in order to coordinate their action if useful, sometimes indicates the parts of the attack front of the divisions on which they will be employed.

He regulates the conditions of use in time and space of the mass-maneuver tank units and ensures that they are protected and supported by powerful artillery.

The organization of antitank defense during the attack is the subject of the Army Corps commander's overall forecast. The second-echelon divisions advantageously deploy their special armament in order to increase the depth of the defense towards the rear.

355. Once the attack has started, the Army Corps commander is informed about its development mainly by the air forces with which he is in direct and permanent contact.

He maneuvers the fires of his artillery, in particular by prescribing, during the fight, powerful concentrations on the points whose conquest or neutralization is of paramount interest.

He directs his reserves, including reserve tank units, to their zones of intervention. He engages them, either by placing them at the disposal of the first-echelon divisions or, if it is a question of a complete division or, in certain cases of a unit of mass-maneuver tanks, by introducing it on a part of the front where he wants to intensify the effort. In other cases, with a view to fueling the battle, he employs the second-echelon divisions to replace the worn-out first-echelon divisions; he then decides whether this operation will be done by passing or by relief.

ARTICLE 5.

COMPLETION OF THE BATTLE.

356. The exploitation of the success or, in case of failure, the preservation of the conquered terrain is organized by the Army Corps commander in accordance with the rules set out in Nos. 240 and following.

CHAPTER II.

THE ARMY CORPS IN THE DEFENSIVE.

357. The Army Corps may be called upon either to hold without thought of retreat a determined position, or to make a withdrawal from action, or to conduct a retreat.

ARTICLE 1.

DEFENSIVE ACTION WITHOUT THOUGHT OF RETREAT.

I. Defense organization and disposition.

358. Having received from the Army commander the indication of the direction to be interdicted as well as the general outline and the lateral limits of the position he must hold, the Army Corps commander, informed by his air force and his cavalry, takes possession of it and pushes beyond the security elements or the outposts, intended to cover the reconnaissance of the terrain, the work of organization, and the progressive adjustment of the defense system.

After having himself acknowledging this position within the framework of the army's defense plan, he makes the decisions relating to the Army Corps defense. These decisions are the subject of a defense plan and relate essentially to the points listed in No. 337 and, in addition, on the following:

— definition of the position of resistance by the traces of its forward limit and its rear limit;

— numbers and missions of the outposts, staking out of their echelons of surveillance and of resistance;

— bases for the organization of the artillery and conditions of opening fire;

— placement and consumption of ammunition;

— general guidelines for the execution of works of defense, signal communications, and communication routes; order of priority.

359. The organization of the terrain is directed according to the indications of No. 264. The Army Corps commander adapts it to all the extent necessary to the requirements of the fight against the enemy armored vehicles (No. 259).

He gives instructions to the Army Corps chief of engineers concerning the distribution among the units of the matériel necessary for the execution of the work.

360. *The outposts* are provided by the first-echelon divisions, each in its zone of action. Their mission in the event of an attack having been defined by the army—withdrawal or defense in place—the Army Corps commander harmonizes, according to the demands of the terrain, the methods of execution on the whole of the front. He sets the points where the outposts of each division will link up with those of its neighbor.

When the distance from the enemy allows it, contact is established and maintained beyond the outposts by the cavalry and, possibly, by security detachments (No. 253), whose composition and mission are determined by the Army Corps commander in accordance with the orders of the army commander (No. 329).

The first-echelon divisions are arranged on the position of resistance (main body) and in the rear (reserves) according to the procedures and with a view to the results prescribed in No. 256.

The Army Corps commander prescribes their reciprocal support and ensures the coordination of their fire plans in the parts adjacent to their zones of action: he controls the value of these plans, in particular for the regions where the penetration of armored vehicles is to be feared.

As long as the attack is not imminent, the reserves of the first-echelon divisions may participate in executing the work, but these troops must be ready to occupy their alert positions at the first signal.

361. The artillery is distributed and echeloned according to the proposals of the Army Corps artillery commander, in view of the various missions defined in Nos. 260 and following.

The groupings at the disposal of the Army Corps commander are more particularly responsible for counterbattery, distant interdiction, and the firing missions on unannounced targets at the request of the air forces;

their action is extended and, if necessary, reinforced by that of the army artillery.

The corps artillery commander prepares *concentrations of fire* on important points of the terrain; he studies the possible intervention of groupings at the disposal of the Army Corps for the benefit of the neighboring Large Units as well as the reciprocal assistance to be requested from them.

362. *The Army Corps reserves* are placed behind the position of resistance according to the requirements of their possible intervention during the battle. They are used for various purposes, as stated in Nos. 268 and following.

They are constituted either by small units taken from the divisions in line, by elements of the general reserves of arms made available to the Army Corps (tank formations in particular), or, more rarely, by an entire division. The intervention of the first two kinds of reserves is generally foreseen within the framework of the defense plan of this or that first-echelon division, with certain constraints imposed by the Army Corps commander. The intervention of an entire division—whether it be a counterattack, the relief of a worn-out division, an entry into line to reinforce the front, or the occupation of a second position or a switch position—requires the Army Corps commander to regulate the conditions of execution himself.

Before the battle, the encampment of the reserves can meet the convenience of the work; but as soon as an enemy attack is imminent, the Army Corps commander has them occupy their alert locations and deploy their antitank armament.

II. Conduct of the battle.

363. The Army Corps commander conducts the battle in his zone of action following the general indications of No. 266 and following.

His air force, whose activity is oriented and regulated by the research plan, constitutes his main intelligence organ; he keeps in direct and permanent contact with their chief.

If, despite the opening of interdiction and counter-preparation fire, the enemy attack succeeds in breaking through, he carefully follows its development in order to give the divisions concerned, in good time, the support of his own artillery and the reinforcements from his reserves.

364. When, having a reserve division at his disposal, he decides to have it counterattack, he himself must ensure its installation, determine its objective and its zone of action, and place all the reinforcements necessary (aircraft, tanks, artillery) under the orders of its leader. In addition, he regulates the use of the artillery that he has directly available for the benefit of this division and requests, if necessary, the assistance of the Army artillery and aviation. Finally, he coordinates the assistance of neighboring troops and organizes the support of the fires to be provided by those who frame the counterattack.

The division commander makes contact as soon as possible with the division commanders who are on line in the region where he is called upon to act. His arrangements should tend to simplicity and vigor, thanks to the procedures set out in No. 270.

Once the counterattack has started, the Army Corps commander takes care to protect it, with the help of his own artillery, against the enemy's immediate reactions or counteroffensives. This protection requires sustained surveillance by his air forces.

ARTICLE 2.

WITHDRAWAL FROM ACTION.

365. The commander of an Army Corps responsible for a withdrawal from action applies the principles and procedures set out in Nos. 273 and 274.

In general, the orders he receives from the Army indicate the echeloned positions that will mark the maneuver and the gain in time that he should ultimately provide. Within this framework, it is up to him to regulate the action of the successive echelons that he will use to slow down the adversary without letting himself get caught. He specifies, for this purpose, the duration of the stop, which will have to be opposed on each of the positions envisaged.

He operates by divisions placed abreast in order to ensure the exercise of command in depth. He decides on the composition of the echelons with regard to the characteristics of the positions they must occupy. He distributes most of the armored vehicles and motorized means at his disposal, organically or by reinforcement, among his divisions, so that they can carry out their successive breaking of contact and combat with the least risk. He adapts to each of them, for the execution of distant fires, a grouping of long-range heavy artillery with the necessary means of aerial observation. He sets the conditions under which the engineer formations will be employed on destructions or their use in the creation of obstacles, the construction of communication routes, etc.

He coordinates ground observation and signal communications over the entire zone of action. He sees to the smooth running of traffic and supplies so that they do not impede the maneuver of the troops and do not attract massive bombardments from enemy aviation.

He obtains continuous information from his air forces and maintains a mobile fire or shock reserve (armored vehicles) to be able to intervene directly in the combat.

ARTICLE 3.

RETREAT AND SPECIAL CASES OF THE DEFENSIVE.

366. For these kinds of operations, the Army Corps commander complies with the rules set out in Title V (No. 275, 277, 291, and 292).

TITLE VIII.

THE INFANTRY DIVISION IN COMBAT.

CHAPTER ONE.

GENERAL.

ARTICLE 1.

COMMAND.

367. The Infantry Division is the *unit of combat.*

The division general commands the various arms and services of his Large Unit.

His essential role is to ***combine the action of the arms;*** only their intimate connection and their simultaneous effort allow results to be obtained at the slightest loss.

The higher command will often have to reinforce the means available organically to the division, especially in artillery, tanks, antitank vehicles, and machine gun battalions.

368. The headquarters of the division essentially comprises:

— a staff;

— infantry, artillery, engineers, and signal communications commanders;

— service chiefs.

The *divisional infantry commander* is at the disposal of the division general for all the missions that the latter deems useful to entrust to him. He acts, in all circumstances, as the delegate of the division general.

In combat, he may be called upon to command a detachment of the division's combined arms forces.

The *divisional artillery commander* exercises, in principle and subject to the reservations expressed in No. 40, the direct command of the whole of the artillery of the division. In accordance with the orders received from the division commander, he forms the various artillery groups, gives them missions, and assigns them areas of action, locations, and observation posts. He coordinates their actions.

The divisional artillery commander is, at the same time, chief of the artillery service of the division.

The *divisional engineer commander* is both a troop commander and a service chief.

The *signal communications commander* exercises the attributions defined by the Instruction on Liaison and Signal Communications.

The *divisional tank commander,* designated when this Large Unit is equipped with tanks, ensures the distribution of the accompanying tank units in the mixed infantry-tank groupings in accordance with the division commander's orders.

During combat, he commands mass-maneuver tank units and the reserve tank units.

The role of the directors or chiefs of services is specified in No. 480.

ARTICLE 2.

EXERCISE OF COMMAND.

369. It is, above all, necessary that the troops receive the necessary orders in time: it is therefore important that the divisional combat orders be **brief.**

Certain arrangements of the order can advantageously be clarified or replaced by summary sketches.

The division commander decides his essential arrangements as often as possible in the presence of the divisional infantry and artillery commanders and, if the situation permits, of the subordinate leaders directly concerned.

He also makes frequent use of warning orders.

Command posts.

370. The division commander establishes his command post according to the indications of time and place that he has received from the higher echelon.

In the offensive, the command post is pushed further forward than in the defensive due to:

— of the lesser width of the zone of action of the Large Unit;

— the more restricted depth of the general disposition;

— a lower probability of deep reactions from the adversary.

This way of proceeding allows, during the advance, a more rapid installation of the signal communications necessary for the exercise of the command.

Taking these considerations into account, the division commander determines the exact location of his command post so as to liaise closely with the division's combat echelon, artillery, and reserves and liaise with the Army Corps commander and the commanders of neighboring divisions.

He also seeks the proximity of observation posts with views of the essential parts of his zone of action so that he can follow the action as closely as possible.

When the situation permits, the division general should temporarily leave his command post to see for himself an accurate account of the situation, to see the terrain of the action, and to make personal contact with subordinate leaders, while remaining in a position to be quickly informed of anything that might require his intervention.

The division commander sets or approves the location of the command posts of the subordinate echelons that he engages directly in his maneuver.

The command posts must have easy access, concealed in particular to the aerial observation and be signposted; they must also be protected against armored vehicles.

The division command post and those of the divisional infantry and artillery, which are generally placed abreast, are moved according to the needs of combat. These moves must be carefully studied in advance and prepared by reconnaissance; during their execution a permanent staff remains at the former command post.

CHAPTER II.

OFFENSIVE COMBAT OF THE DIVISION.

371. The resolution to attack results from the mission received: it is up to the division commander to conquer the successive objectives fixed by the Army Corps commander and to achieve a judicious combination of arms and their fires.

The division's maneuver varies according to its place in the disposition (a division placed abreast of other troops, on the wing, exceptionally isolated), the importance of the security elements operating in the zone, and depending on whether it is acting on open ground, against a moving or stopped enemy, or if it must attack a stabilized or fortified front.

The division commander establishes the main lines of his plan of maneuver as soon as possible; he settles the details, usually during the establishment of contact. At this time, the intelligence will probably be incomplete: the desire to identify it too much could lead him to make his decision too late, wearing himself out prematurely, and playing into the enemy's hands.

ARTICLE 1.

APPROACH MARCH.

372. The division carries out its approach march according to the general rules prescribed in No. 290 and within the framework fixed by the Army Corps (No. 339).

The division's approach march disposition is echeloned in depth and width; it is articulated, in principle, in combined arms columns or groupings, preceded by advance guards. It comprises successively:

— *distant security elements* (reconnaissance battalion). Distant security will often result from arrangements made by higher echelons: the reconnaissance battalion will then liaise with the army or Army Corps elements, ready to reinforce them or to take on their mission of intelligence and covering.

If the division alone must provide for its own distant security, this mission falls to the reconnaissance battalion, reinforced in particular with motorized elements and antitank weapons (No. 138).

The reconnaissance battalion, operating at distances related to the maneuver of the division, then searches the critical directions of the zone of action for the intelligence requested by the division general. It stands ready, as soon as contact has been established, to control the most favorable lines for an economical defense, particularly against armored vehicles;

— *close security elements* (advance guards, flank guards) capable, if necessary, of combing the zone of action in all its width and which cover the main body at a distance equivalent to the average range of an adversary's field gun (No. 144);

— *the main body* whose center of gravity is placed on the direction assigned to the division.

Approach march at night.

A night approach march can only be considered if the division is assured of carrying out its march without fighting.

This imperative condition is satisfied when the approach march is carried out sheltered from a pre-existing front or when the distant security elements have been able to have the necessary means to install before nightfall, effective coverage of the division's approach zone.

The main body of the division moves preceded, on each route of the zone of action, by advance guards reduced to small elements capable of interdicting the communication route network.

Behind these advance guards, the main body, split into march groupings, moves in columns, using all the distinct routes available (No. 170).

In the immediate vicinity of an adversary in position, the night approach march must be carried out under the protection of the advance guards which will have occupied, before nightfall, the points of the terrain necessary to cover the movement.

Approach march during the day.

The division makes its approach march by bounds, informed and covered by aviation and security elements.

To this end, within the framework fixed by the Army Corps, the division general regulates the advance according to the compartments of terrain favorable to his possible maneuver. He deduces the paths and itineraries that lead to it; he specifies the bounds of the main body and, consequently, those of the security elements. If the configuration of the terrain allows it, it will be advantageous to assign to these bounds limits favorable to the rapid organization of the defense against armored vehicles.

The disposition moves first by looking for concealment and cover and by taking the most appropriate formations to accelerate the movement as well as to protect the troops against the effects of armored vehicles, aviation, and enemy artillery.

As one gets closer to the enemy, the need arises to put the forces in hand for combat. For this purpose, the division general progressively makes the articulated formations known to the advance guards and then to the main body. Units are echeloned in width and depth so that they can move in order and, if necessary, deploy quickly to form a front.

During the approach march, the artillery, dispatched by the division general, establishes its liaison with the infantry, whose advance it carefully follows. It stands ready to intervene very quickly, gradually putting its batteries in position, moving, if necessary, by echelons to support the advance guards with part of its matériel.

The division general closely follows the progression of the advance guards.

ARTICLE 2.

ESTABLISHMENT OF CONTACT.

373. Contact is established by the division under the conditions prescribed in No. 345.

In front of an enemy in position, the advance continues until the advance guards have penetrated far enough into the enemy disposition to encounter a continuous line of fire that they cannot break by their own means.

If *an enemy is on an offensive march,* the advance stops as soon as the advance guards arrive at the height of the position on which they must, according to the orders of the Army Corps commander, install themselves defensively to stop the opponent.

In both cases, contact should create a front from which attacking units can make their security arrangements.

The division commander, from the first contacts of the reconnaissance battalion, coordinates the movement of the advance guards with that of the artillery and the main bodies; he moves up to the combat echelon of the advance guards, at points favorable to observation.

The artillery continues its advance by echelons and stands ready, if necessary, to support the infantry with all means.

It remains, in principle, under the orders of the division general. But the latter must not hesitate, if the structure of the terrain or the extent of the zone of action makes it useful, to agree to all the necessary decentralization of the artillery command to obtain more effective and faster support from the advance guards (No. 40).

The main body of the infantry continued its march in bounds, keeping a sufficient distance not to feel too soon the fire of the enemy artillery nor the reactions incurred by the advance guards.

The division commander takes the combat in hand as soon as the advance guards, spread over a wide front and without depth, are stopped in the presence of serious resistance (case of an enemy in position) or subdued, on the position that they hold, under vigorous pressure from the advanced elements of the adversary (case of an enemy on the offensive march).

At this moment, the role of the advance guards ceases; the division commander, with all the means at his disposal, except those reserved by the higher echelon, puts himself in a position to conquer the fixed objective (case of an enemy in position) or to hold on to the designated position (case of an enemy on the offensive march).

Once established, contact must be maintained day and night and, if lost, resumed as soon as possible to avoid any surprise, particularly that resulting from an attack launched in a vacuum or at a great distance from the enemy.

374. During the approach march and the establishment of contact, liaisons and signal communications must be organized in such a way that:

— the division general can receive intelligence from the reconnaissance battalion and the advance guards without delay;

— the advance guards are in contact with each other and with the reconnaissance battalion, and are linked with the direct support artillery;

— the division general can quickly send orders for the engagement or halt to the various elements of the main body.

These relations are ensured first of all by rapid agents of signal communications, then by telephone.

The use of radiotelegraphy is subject to the restrictions essential to maintaining the secrecy of the disposition.

ARTICLE 3.

ENGAGEMENT.

375. The engagement being the first act of the attack (No. 221), the commander who resolves to attack in principle orders the engagement.

It follows that, generally speaking, the division commander makes this decision only to direct the action prescribed to his Large Unit in the engagement of the Army Corps.

This action by the division, whether or not in conjunction with the other divisions of the Army Corps, differs from the attack carried out in battle only in terms of its limited purpose, front, and range, as well as its relative isolation.

Because of the division's weak means in this phase of combat, it will include using a limited number of infantry, for which it will be important to ensure preparation and support are as effective as possible, framed by fire and covered by the terrain. All the artillery available to the division will be called upon; tank support will be particularly sought after.

If the enemy yields, efforts will be made to widen the breach as soon as possible and to continue the advance in the direction fixed by the Army Corps.

376. The organization of signal communications is characterized by a reinforcement of means for the benefit of the units (particularly the artillery) in charge of the engagement.

The division's telephone network is pushed closer to the command posts of these units.

The use of radiotelegraphy no longer has any restrictions, at least for engaged units.

ARTICLE 4.

ATTACKS.

377. The division will generally be reinforced for the attack.

It may receive, depending on its mission and the availability of the command:

— short-range light or heavy *artillery,* exceptionally long-range heavy artillery, sometimes the artillery of a reserve division with certain constraints of use;

— *tank units* of various types;

— *antitank units.*

Possibly:

— *machine gun battalions*;

— *engineer elements*, of a reserve division or of the Army Corps with certain constraints of use;

— *air forces* (balloon, squadron, etc.).

Plan of attack.

378. The plan of attack aims to coordinate the division's use of organic or reinforcement means to conquer the objective set by the Army Corps in a given direction.

If, because of its remoteness or its width, the objective assigned by the Army Corps cannot be taken at a single momentum, the division commander may be brought in, with a view to combining the action of arms, to set intermediate objectives determined according to the means of support (artillery and armored vehicles) at his disposal, the resistance of the enemy, and the nature of the terrain.

With the success obtained and the first objective having been reached, the division commander goes on to attack a more distant objective so that the maneuver takes place without loss of time.

A division attack order must convey the general situation and the mission of the division, the idea of the maneuver, the direction, the objectives, the disposition, and the missions of the infantry regiments and the various arms.

The order also gives prescriptions concerning:

— the organization of defense against armored vehicles during the attack;

— the conditions for deploying the attack troops, the mode of preparation, the zero hour;

— liaisons with neighboring divisions;

— the location of command posts and the organization of signal communications;

— the functioning of communication routes, supplies, and evacuations.

Disposition.

379. The division's combat system is articulated by the division general, who distributes his forces in width and depth in order to ensure:

— on the one hand, the desired initial density on the front of attack;

— on the other hand, the echelonment of the means is likely to guarantee the development of efforts in time and space.

The combat disposition of the *division* includes the combat echelon, the artillery, and the division's reserves.

380. The *combat echelon* is made up of:

— the necessary and sufficient infantry and accompanying tank units to obtain superiority of fire from the outset, avoiding any excessive density that would increase losses without profit;

— possibly mass-maneuver tanks intended to precede the infantry and accompanying tank units on their successive objectives (No. 38 and 230).

The infantry and accompanying tank units are brought together in mixed groupings juxtaposed or separated by intervals. These groupings receive from the division commander a leader and a mission.

381. All the *artillery* is articulated at the request of the infantry disposition while remaining capable of contributing to the concentrations of fire that the circumstances of the combat may require.

For this purpose, the division commander generally distributes his artillery into two so-called detachments, one for *direct support* and the other for *general action* (No. 235).

The *direct support detachment* is divided, according to the missions and the terrain, into *groupings* corresponding in principle to the first-echelon infantry regiments, whether or not these are equipped with accompanying tanks. When mass-maneuver tanks operate in front of the mixed infantry-accompanying tank groupings, these are put in contact with direct-support groupings acting for their benefit and are operated by them under the conditions provided for in No. 235 for support groupings in general.

Direct-support groupings provide, in principle, close-support fire, either according to a pre-established plan, or according to requests sent by the infantry or tanks to which they are directly attached.

The *general-action detachment* allows the division commander to make his action felt during combat, particularly through concentrations of fire. It usually provides protective fire but can also reinforce close-support fire.

In general, the artillery disposition will be pushed as far forward as possible; the direct-support groupings will, in principle, be placed in the axis of the attacks; the search for enfilade effects may lead to the establishment, after agreement, of certain batteries in the zone of neighboring divisions.

382. In particularly rugged, covered, or broken terrain presenting clearly defined attack corridors, it may, *in exceptional cases*, be advantageous to use tactical groupings (see general definitions, page 23).

383. *Division reserves* consist of uncommitted infantry and possibly tanks. The division commander may be required to keep an entire regiment in reserve; this arrangement is generally necessary when the division is on a wing.

The reserves are articulated and arranged with a view to their probable use: attack support, exploitation maneuver, possibly extension of the line of battle, or protection of an open flank. In some cases, they can, while waiting to be engaged and in order to ensure the preservation of the conquered terrain, reconnoiter and organize the positions reached during the advance.

Setting up the disposition.

384. *The attacking infantry must face their objective in a clear direction, and be well supported by fire and protected on its flanks.* The placement aims to achieve these initial conditions.

It will always be advantageous to bring the leading elements to a *base of departure,* consisting as much as possible of the last cover before the positions occupied by the enemy.

A base of departure close to the enemy makes it possible to approach the opposing position in good conditions; too close can, however, hamper the artillery's action, compromise the security of the preparations, and deprive the attack of the benefit of surprise.

In mobile warfare, setting infantry in place is often a delicate operation; on open terrain, in particular, it can only be carried out at night under the shelter of a solid front and under the protection of a vigilant counterbattery.

If setting up in the daytime is required, the division commander endeavors to obtain from the higher command control of the air and the blinding of ground observation posts.

The accompanying tank units are placed as close as possible to the infantry units, unless their speed allows them to take their initial position at a distance from the base of departure that puts them out of reach of enemy counter-preparations.

Defense against armored vehicles.

385. Defense against armored vehicles aims to protect the division's posture, its flanks, and its rear during all phases of the attack.

It rests:

— on the furthest possible search for and rapid exploitation of intelligence;

— on the combination of natural or artificial obstacles, with concentrations of fire (aircraft, artillery, organic or reinforcement antitank weapons, mines, etc.);

— on the intervention of the tanks assigned to the attack, particularly on reserve tanks.

Defense against armored vehicles is organized within the framework of the division.

It rests on:

— the division's artillery;

— divisional antitank units or those assigned as reinforcements to the division;

— antitank weapons that the units have organically;

— and possibly, the tanks that could have been provided for the attack.

The disposition of the defense against armored vehicles is characterized by its ***depth.***

As a general rule, it successively comprises:

— *a mobile echelon* advancing by bounds in the wake of the infantry so as to provide it with the most constant support possible from its fires;

— *a second echelon* intended to counterattack the enemy vehicles that may have crossed the preceding echelon and thus cover the artillery, the command posts, and the rear;

— *a barrage constituted* mainly by the organic artillery of the Large Unit and by its own antitank weapons.

When the situation requires it, this disposition is supplemented by the establishment of barrages intended to ensure the coverage of the flanks.

It is up to the division general to coordinate the use of various means in order to organize a coherent and deep antitank fire system and to establish the *anti-armored vehicle plan* accordingly.

Preparation for the attack.

386. Artillery preparation is carried out under the conditions prescribed in Nos. 232 and 399.

The use of tanks for the attack does not eliminate artillery preparation: the prior disorganization of the adversary's antitank system is, in fact, essential. In addition, when the opposing defense relies on obstacles that are impassable to tanks or organizations that they cannot remove by their own means, it will be necessary to open the way for them by using a prior action of other arms comprising, in particular, the use of powerful artillery.

Execution of the attack.

387. Under the cover of intense action from the available artillery and automatic weapons bases of fire, the mixed infantry-accompanying tank groupings debouch at a fixed time or at an agreed signal (No. 233), with the accompanying tanks operating in the ranks of the infantry or immediately preceding it.

The infantry executes by bounds, towards the objectives, an advance from cover to cover or from shelter to shelter constantly prepared or supported by fire; the size of the bounds depends on the need to maintain throughout their duration the neutralization of enemy fires acquired by the superiority of the fires of the attack. The infantry squeezes as close as possible to the projectiles of its artillery, operates in close liaison with the accompanying tanks, and adapts its maneuvers to the terrain. It exploits, as soon as they appear, the effects of neutralization and destruction obtained by the artillery and tanks. In return, it provides the latter with the support of its fires against antitank weapons.

During its advance, the infantry endeavors to quickly resolve the incidents of the combat with its organic means and within the framework of the close-support fires initially envisaged. If it does not succeed, it asks its direct support artillery to neutralize the resistance that opposes its advance. If this support proves to be insufficient, the division general intervenes, then and can grant the assistance of the general-action artillery, which always requires a considerable delay.

The mixed groupings thus gain ground by crushing the enemy under their fire; the small units facilitate the advance of their neighbors by their own movement, force the enemy to hide, and, if necessary, dislodge them from their positions by assaults starting at a good distance and at the right moment.

The attack takes on the character of a full assault, at an agreed time or signal, only in front of a position or an objective that presents a continuous front, and starts from a base of departure substantially parallel and close to the front.

388. As soon as an objective is captured, the attacking troops clear it and settle firmly under the protection of the artillery, avoiding overcrowding. The units are put back in order and resupplied, the fires of infantry, antitank weapons, and artillery are adjusted, the signal communications established, the liaisons established, and the terrain is organized in a summary way. If circumstances permit, after agreement with the direct-support artillery, contact patrols are launched immediately. Elements are kept on the spot until the conquest of the next objective and are used, in particular, to constitute new bases of fire; they are advantageously equipped with antitank weapons. Reserve units can participate in this mission.

The attack on the next objective is immediately undertaken. According to its nature, that of the terrain which precedes it, and according to the effort already provided, the attack is continued by the same units, or is taken again only after reorganization of the disposition and engagement of fresh troops; reliefs and the passing of lines are then regulated so as not to compromise the solidity of the front and to avoid the losses which a momentary accumulation of troops could cause.

As soon as it is useful, the division commander pushes his artillery forward; the latter executes its movements in such a way as to be constantly in a position to give the infantry effective support.

The division commander regulates the movement of his reserves and ensures their intervention in good time to guarantee the continuity of the effort and to prevent the enemy from recovering. As he spends his reserves, he seeks to replenish availabilities with units withdrawn from fire.

389. The attack continues in this way until the final objective.

Nighttime is used to rectify the end-of-day arrangement and improve the fire plan, to reconstitute, resupply, and relieve the engaged units, to organize the conquered position against the armored vehicles as a matter of first priority, and finally, to restore the signal communications and arrange the communication routes. It can also be used to cross areas that are difficult to cross during the day and thus put the troops in a position to continue their offensive the next day under advantageous conditions.

Maintaining contact is one of the division commander's essential concerns. For this purpose, as soon as an objective is reached and when the situation allows it, he can seek contact by using a strong reconnaissance, subject to coordinating his action with that of the artillery.

390. The division commander directs and combines the fires (No. 396). He regulates, in particular, the use of the fires of the Army Corps heavy artillery he may have for the attack, and finally, he sees to the implementation of the defense plan against armored vehicles.

The aviation informs and contributes to ensure the liaisons with the artillery command.

391. When the Army Corps commander places at the disposal of the division units of mass-maneuver tank intended to constitute the first echelon of the attack, the division general sets their mission and their zone of action and, in particular, coordinates their advance with that of the artillery fire.

These deeply echeloned tank units strive to neutralize enemy weapons in favor of the accompanying infantry or mixed infantry-tank groupings which they precede as widely as the compartmentalization of the terrain allows.

They supplement the artillery support of the infantry', but demand from the latter serious protection during the action and often preparation for their entry into line.

Organization of signal communications.

392. During the attack's preparation, a solid telephone network capable of rapid expansion is established as a first priority.

During the attack, the establishment and functioning of signal communications presents particular difficulties because of the frequent movement of the command posts of small units.

Radio networks are implemented without restriction.

With regard to the other means (telephone, optical, signal communications agents), relations between the division's command post and the command posts of the directly subordinate units can only be established without intermediaries for attacks of limited magnitude. In other cases, intermediate signal communications centers will often be necessary.

All efforts must focus on organizing optical signal communications, maintaining the established telephone network in good condition, extending it as needed, and using the circuits already established by the subordinate units.

ARTICLE 5.

COMPLETION OF THE BATTLE.

393. Any success obtained is immediately exploited to the full, either by virtue of command orders or directly on the initiative of subordinates. To orient the latter and ensure the liaison of arms, it is a good idea to include in the orders the main operating maneuvers planned and to echelon the disposition accordingly.

The exploitation develops:

— in width to bring down, through a combination of front and flank actions, the parts of the enemy position extended beyond by the advance made and thus widen the breach;

— in depth in the assigned direction, covering the flanks, but without adjusting to the halted units; the continuous push of the elements that can advance is the best means of freeing up neighboring units.

Exploitation is in principle the responsibility of first-echelon troops in combat condition, reinforced as far as possible by fresh units and by tanks, the use of which is particularly prescribed in this phase of combat; the tested elements, maintained on the conquered position, give to the troops which pass beyond them the support of their fires; they organize this position and go into reserve.

The exploitation maneuver continues in successive bounds until the final objective assigned to the division is reached. The division commander wants to ensure that during this maneuver, the decentralized artillery, to the fullest extent necessary, is always in a position to support the infantry (No. 40). The divisional engineer commander restores the signal communications necessary for the forward thrust of the artillery and the arrival of supplies within the general framework drawn up by the higher echelon.

394. If the beaten enemy withdraws in disorder, the *pursuit* begins. Speed then becomes the essential factor in the maneuver: each unit exploits to the full, without adjusting to the others, any success obtained.

Begun by the first-line troops, the pursuit uses the cavalry, armored vehicles (No. 38), and motorized units as soon as possible. Aviation informs the advance and intervenes on the ground.

The division commander constitutes, as soon as he can, strong advance guards well equipped with armored vehicles and artillery and includes engineer detachments equipped with the necessary crossing equipment if these advance guards have to encounter terrain breaks at short notice. Their mission consists, on axes fixed by the Army Corps, of quickly outflanking local resistance and, in particular, getting ahead of the adversary's delaying elements on the terrain breaks.

The main body of the division, artillery in the lead and arranged in march groupings, follows, ready to intervene.

395. If the attack has not succeeded, the infantry clings to the ground until the command has mounted a new attack or until the advance of neighboring units has allowed a resumption of movement. It strives to organize its fire as soon as possible, settles on the ground, and organizes its defense against armored vehicles; it stakes out its front to indicate to the air force its exact situation, it is covered by obstacles, and replenishes its arrangements.

The artillery sets up its stopping and counter-preparation fires.

The attack resumes, either on the order of the higher command or on the initiative of the local command.

Before resuming the attack, it is important to first determine the main cause of the failure in order to organize the new operation accordingly.

In all cases, artillery preparation is essential, and the use of superior matériel means is essential.

ARTICLE 6.

MANEUVER OF FIRES.

396. At all echelons, the command has the imperative and constant concern to prepare, conduct, and combine all the fires at its disposal *in order to support the infantry and seek the destruction of the tanks and antitank vehicles of the enemy.*

He also takes care to ensure the protection of the disposition and his rear against the action of the opposing air forces.

I. Artillery fires.

397. The division commander directs the action of his artillery during all phases of combat.

The division's artillery prepares (No. 232) and supports the attack (No. 236); it also executes the fires that may be prescribed to it by the Army Corps, both before and during the attack (interdiction fires, counterbattery fires, etc.).

The modalities of the artillery's action in the attack vary according to whether or not it is equipped with tanks and according to the mission assigned to it (No. 236).

398. When the attack does not have tanks, close-support fire takes on primary importance and must be adjusted as close as possible to the first elements of the infantry.

When the attack has tanks, it becomes necessary to give them, between the first infantry elements and the shortest artillery fires, ***a field of action whose depth depends on the terrain, the characteristics of the tanks used, and their maneuvering in relation to the infantry.***

— if it is only *accompanying tanks*, the system of close support and protection fire remains with, however, an extension of the short limit of close-support fire to allow tanks to operate.

— if the disposition also includes an echelon of *mass-maneuver tanks,* preceding the accompanying infantry-tank system, the artillery devotes itself to the protection of the first echelon of tanks, with which certain artillery groupings are placed in direct liaison to give their fires all the necessary flexibility.

But, in the latter case, the prior organization of the artillery command and its adaptation to the infantry must make it possible to restore, as soon as possible, the complete system of fires (close support and protection) for the benefit of the first elements of infantry if they cease to be covered by tanks.

Preparation Fire.

399. Preparation fires include *fires as violent as possible, the duration of which varies from a few minutes to several hours.*

Preparation fire is mainly directed at enemy infantry positions, accessory defenses covering them, and antitank weapons detected by ground or aerial observation.

Even when these positions could not be determined with precision, the artillery could still prepare the attack by taking under fire the objectives to be removed, the parts of the terrain from which the enemy would have action on the attack, such as the likely assembly areas of reserves and enemy batteries, and the suspected locations of antitank weapons.

Close-support fire.

400. Close-support fire generally consists of a bombardment of the closest targets, this bombardment being maintained until the moment when the progress of the attack makes it necessary to lengthen the fire.

Close-support fires are intensified in zones where the infantry makes the effort, adapting to the terrain's organizations and obstacles and taking on any antitank weapons that are revealed.

They can, when the combat echelon has come within close range of an enemy whose organization could not be recognized in all its details, and when the front assigned to the division as well as the quantity of artillery and ammunition at its disposal allow it, take the form of a rolling barrage followed as closely as possible by the first line and settling in front of it at the stops provided for in the timetable.

Often the rolling barrage will only be used at the start of the attack to support the movement of the infantry up to a first objective, clearly inscribed on the terrain and necessarily including the first enemy resistances.

When the attack has made some progress, the direct support artillery moves itself, if necessary, to ensure its mission without interruption. At this time, the decrease in the number of batteries in a position to fire and the difficulties of supply will lead to the use of rapid concentrations executed by matériel ready to intervene.

In such circumstances, the momentary weakening of artillery actions may lead, if the situation permits, to the use of available tanks.

Protective fire.

401. Protective fire prolongs the action of close-support fire (No. 235).

They take hold of escaped objectives that appear during the action, particularly distant antitank weapons and counterattacking troops, either during their assembly or at the time of their debouchment.

They blind probable enemy observation posts and suspected locations of antitank weapons.

When, during the advance, the command is uncertain about the exact situation of its infantry, it strives to come to its aid by giving all possible development to protective fire.

Coordination of artillery fire with infantry movements.

402. The coordination of artillery fire with infantry movements is an essential condition for success. Most often, and almost necessarily, even if it is a question of attacking an organized position, this coordination is regulated by a schedule or at least planned in detail at the start and for a certain depth of attack. It can then only be deeply varied thanks to marked or at least planned halts on objectives specified in advance.

As a result, this coordination is essentially based on the proper functioning of the infantry-artillery liaison.

The realization of this liaison includes:

— frequent contacts between the artillery commanders and the infantry commanders. These contacts are greatly facilitated by the proximity and, even better, by the juxtaposition of command posts (artillery-infantry), each time that the exercise of command does not suffer from it, and by the meeting at the same observation posts of the representatives of the two weapons;

— the use of liaison detachments: any group and, if the situation so requires, each direct support group provides the infantry formation it supports with a liaison detachment responsible for informing, on the one hand, the infantry on the support that can give him the artillery and, on the other hand, the artillery on the needs of the infantry;

— the forward thrust, constant and as rapid as possible, of the artillery observation posts (visual link).

II. Infantry fires.

403. The infantry gives its fires all possible power.

It combines them with the action of the tanks to prepare and accompany attacks itself, cover its flanks, protect the tanks against enemy vehicles firing at short distance, defend itself against enemy armored vehicles, repel counterattacks, ensure occupation of conquered land.

It moves its arms in echelons to ensure constant support for the advance and support the attacking troops to the end.

She searches for all the possibilities of slung fire or reverse action on enemy resistance that she has been able to outflank.

In the event that an echelon of mass-maneuver tanks precedes the accompanying infantry-tank system, the infantry takes part in its protection by its fire.

III. Air defense fires.

404. Air cover for the attack is assured under the conditions laid down in Nos. 301 et seq. All the units of the division ensure, moreover, by the implementation of their organic means, their Local air security (No. 156).

ARTICLE 7.

ATTACK FROM A STABILIZED FRONT.

405. The attack on a stabilized front is prepared and conducted in the general conduct provided for in No. 244.

The prescriptions relating to the preparation and execution of the attack are brought together in sufficiently detailed orders to contain all the necessary details, in particular with regard to the use of artillery; these orders will be usefully supported by sketches.

The attack troops are articulated in such a way as to be able to ensure the clearing of the defense shelters without their progress being slowed down. Detachments designated in advance and provided with special weapons are responsible for this clearing; small elements of engineers can be added to them for the search for time-delayed mines. The levies to be expected as a result will generally lead to a reduction in the depth of the terrain that a unit of a given size is likely to conquer. This should be taken into account for overtaking and relief.

The use of powerful tanks for the attack of a stabilized front provides, on the other hand, the advantage of aiming at distant objectives, on the condition that these vehicles have been put in a position to cross the most formidable obstacles and enemy organizations without hindrance. This last result is obtained either by using an artillery preparation connected to this objective or by engaging these tanks only beyond the resistances that they are unable to overcome.

406. In the presence of a stabilized front, it is particularly necessary to carry out small operations aimed at clearing the terrain in front of the enemy position and securing the possession of certain important points of this terrain.

These operations, like those aimed simply at capturing prisoners, take on the character of more or less important raids.

Coups de main are, in general, attacks with limited objectives whose success depends mainly on surprise. Their execution involves the use of artillery, and sometimes aviation, set up to open the way for the infantry and to cover it during the occupation of the conquered ground.

In execution, it is generally advisable to use protective artillery fire surrounding the defenders to remove any possibility of reinforcement or withdrawal.

407. During an attack on a stabilized front, timetables are usually used to coordinate artillery fire and infantry movement.

However, compliance with these schedules becomes uncertain when they apply to an attack that is developing at a great depth or extending over a long period.

To keep the progress of the projectiles of the artillery and that of the attacking troops in harmony, it is important to give the infantry time to overcome the difficulties resulting from the defensive works and the state of the terrain. For this purpose, it is advisable to provide halts of suitable duration on successive lines that are very clear, easy to recognize, and lend themselves to subsequent debouchments. In addition, a small number of clear and simple conventions, based on the use of optical signal communications, air liaisons, and radio communications, will make it possible to carry out, at the request of the combat echelon, the necessary modifications of the schedule in certain urgent cases

408. During the attack on a stabilized front, the organization of signal communications is designed as stated in No. 392.

However, within the division, it will be possible to establish direct relations, as in a small-scale attack, and not resort to intermediate signal communications centers.

CHAPTER III.

DEFENSIVE COMBAT OF THE DIVISION.

409. The division may be called upon to hold a specific position without thought of retreat, to make a withdrawal from action, or to conduct a retreat.

ARTICLE 1.

DEFENSE WITHOUT THOUGHT OF RETREAT.

410. In this situation, the division commander has, in his zone of action, the duty to ensure, by all means, the integrity of the position of resistance established by the Army Corps commander.

To this end, it involves:

— *fires,* to break up and then stop the attack;

— *movement,* in order to maintain or re-establish the integrity of the position by the intervention of the reserves.

It will always be in his interest, moreover, to ensure the benefit of *surprise* in order to leave the enemy as long as possible in ignorance of the exact location of the position and of the strength of his occupation.

I. The defensive position.

411. The defensive position of the division (position of resistance and system of outposts) is conceived according to the general rules set out in Nos. 249 and following.

The position of resistance and, therefore, the main line of resistance must, as a rule, be covered by a natural or artificial *obstacle.*

The stop line must also rest as much as possible on an obstacle to ward off deep incursions by armored vehicles. Its layout is also established to facilitate flank actions favorable to counterattacks. Finally, it must provide cover to the artillery.

The depth of the position of resistance (No. 249) must give the defense the possibility:

— avoid the accumulation of resources and thus reduce the vulnerability of the troops;

— to carry out an echelonment of its weapons in order to compel the enemy to make successive efforts and to allow the eventual re-establishment of the barrage of fire.

It is all the more necessary when the defensive position's organization is more basic or when the obstacle covering it offers less value.

It will often be conditioned, moreover, by the interest that attaches to covering the stop line with an obstacle.

412. The position of resistance and the general conditions of the installation of the system of outposts are defined by the Army Corps commander (No. 358), who also indicates the directions and essential parts of the terrain to be held in the zone of the division.

The division commander, therefore, determines the points of force of the organization to be carried out and decides on the detailed layout of the main line of resistance according to the band of terrain on which it has been decided to apply the general barrage (No. 251).

The outpost system is organized according to the indications of No. 252.

When the role of the outposts is not fixed by the higher command, the division commander specifies it, taking into account, in particular, the degree of advancement of the organization of the terrain on the position.

He takes care to practice in this respect, a strict *economy of forces,* with a view to devoting the maximum of his means to the defense of the position of resistance. In certain favorable circumstances, in particular when this position is established behind a major natural obstacle, the role of the outposts may be reduced to simple surveillance.

II. Defense organization and preparation.

413. After a rapid reconnoitering of the terrain, the division commander issues an order intended to ward off an immediate attack; the troops are placed on the position of resistance, close to their combat locations and in conditions favorable to the execution of the work; they are covered by outposts and take protective measures against aircraft and armored vehicles.

Defense plan.

414. The divisional *defense plan* (No. 254) defines:

— the general situation and the possibilities of the enemy;

— the mission of the division and neighboring units;

— the idea of maneuver (essential directions to deny and points to keep);

— the layout of the position of resistance (main line of resistance, stop line, essential switch trenches);

— the outpost system (staff, missions, locations of the surveillance and resistance echelons);

— the system of defense against armored vehicles;

— the disposition (combat echelon, artillery, reserves);

— the mission of the subordinate authorities (subsector, artillery, engineers, air forces);

— use of reserves;

— the works to be carried out with an indication of their order of priority and the distribution of supplies of ammunition and matériel;

— the organization of the liaison;

— the location of the main command posts;

— the organization of signal communications;

— the organization of the communication routes, supplies and evacuations.

Disposition.

415. In the defensive, the division generally fights in regiments placed abreast; at all levels the command is organized in *depth.*

The division commander sets, according to the orders of the Army Corps, the distribution and the missions of the troops of his Large Unit.

The combat disposition includes the combat echelon, the artillery, and the division's reserves.

416. *The combat echelon,* essentially made up of the majority of the infantry, is placed on the position of resistance, echeloned in its depth.

The division commander sets the important points to hold and even, if necessary, the location of certain *centers of resistance* and essential *strong points.* He determines the number of battalions to put in line. He then subdivides the sector of the division into subsectors, generally corresponding to a regiment.

The commander of a subsector determines the distribution of his battalions between the various points of force of his subsector. The intervals, carefully covered in particular by the flanking fires of the centers of resistance and the strong points that surround them, are, if possible, barred with obstacles; reduced numbers are allocated to them to make any infiltration of the enemy under cover of night, fog, and smoke impossible.

Each battalion organizes itself in its quarter by securing the necessary reserves to eventually restore the continuity of fire and launch immediate counterattacks.

The outposts are provided, in principle, in each subsector, by the units belonging to the battalions in charge of the defense of the position of resistance. There is always an interest in providing them with ample means of signal communications.

417. In general, the division commander divides, as in the offensive (No. 381), his artillery into a fraction called *direct support* and into another called *general action.* He then distributes the direct support artillery among the different subsectors according to his idea of maneuver.

The artillery remains entirely, in principle, under the orders of the divisional artillery commander, subject to granting all the necessary decentralization when, in particular, the extent and nature of the front to be defended make it useful.

The infantry-artillery liaison is organized following the same rules as in the offensive (No. 402).

The artillery is deployed as described in No. 257. Whatever their sites, the batteries provide close defense against the armored vehicles in the first place.

418. *Division reserves* consist primarily of uncommitted infantry and, possibly, tanks.

The division command most often builds up its reserves by taking battalions from the regiments of the combat echelon.

The reserves occupy positions behind the position of resistance that allows them either to reinforce the line of combat, execute counterattacks corresponding to the most probable eventualities, or, in the event of an enemy penetration, adopt a defensive attitude on the spot and without delay.

If the division has a wing, the reserves take part in the protection of the open flank, in particular against armored vehicles. The divisional reserves then should include a full infantry regiment.

In this case, the reconnaissance battalion is carried on the flank to inform or cover it.

Organization of the terrain.

419. The organization of the terrain essentially includes the placement of fire weapons, the reinforcement or creation of obstacles, in particular antitank obstacles, the installation of command units, and the arrangement of communication routes and signal communications.

It is the subject of an *organization plan.*

It is conducted in a progressive manner so as to be able at any time to lend itself to good use of the defense troops. The prior camouflage of the work is of paramount importance.

The organization plan is possibly supplemented by a *destruction plan in* accordance with the mission assigned, in this respect, to the division by the higher echelons.

When an enemy attack seems imminent, work is limited to first-priority items (No. 265).

Defense against armored vehicles.

420. Defense against armored vehicles in the defensive aims at the same goal, relies on the same elements, and uses the same means as in the offensive (No. 385).

It is organized within the framework of the division.

The defense disposition against armored vehicles develops over the entire ***depth*** of the defensive position.

As a general rule, it successively comprises:

— *an echelon* established by the outpost units;

— *a main antitank barrage* which it will always be advantageous to make coinciding, as much as possible, with the general barrage;

— *interior barrages* set by antitank weapons echeloned in the corridors most favorable to the penetration of enemy tanks;

— *a barrage* established at the level of the stop line and intended to cover the artillery, the command posts, and the rear;

— *a rear barrage* consisting mainly of organic artillery and its own antitank weapons.

These barriers may not be continuous; it is enough to deny the zones not covered by obstacles, that is to say the weak points of the defense.

The protection of the flanks is also ensured by using special barrages, established on the switch positions set by the higher command.

Tank units, possibly assigned to the division and maintained by it in reserve, can also be used to hinder the action of enemy armored vehicles. The intervention of these units will generally take the form of counterattacks for which the artillery, infantry, and antitank weapons, in a position to act, will provide support, protection, and cover for the flanks, in the framework of the projections based on the most probable directions of enemy tank attacks.

The entire defense system is the subject of the *plan of defense against armored vehicles.*

The complete organization of a defensive position against armored vehicles requires long delays and significant matériel resources; it must, therefore, be undertaken without delay and pursued continuously.

Organization of signal communications.

421. The organization of the signal communications in the defensive presents certain particularities concerning the use of the telephone.

Normally the division command post should be linked by telephone with the subsector commanders, the infantry and artillery commanders, and with the higher commander (possibly with the balloon and the airfield).

In addition, the telephone is, as far as possible, pushed to the infantry and artillery battalion, and even company and battery command posts as well as to important observation posts, all precautions being taken to avoid interception by enemy eavesdropping.

There is value in linking the division command post by telephone to the outposts through the subsector commanders.

III. Conduct of the combat.

422. When the necessary space is available, the command may be required to push delaying detachments in front of the enemy under the conditions prescribed in No. 253.

These detachments strive to delay the advance of the adversary's forward elements, forcing them to deploy and then withdraw under cover of the night and the terrain.

Combat of the outposts.

423. When the outposts have received a mission to resist on the spot under the conditions specified by the command (No. 252), their combat presents characteristics analogous to that for the defense of the position. Resting on organized terrain, supported by a number of heavy infantry weapons from the position of resistance and designated artillery units, covering their approaches and gaps, the outpost units use their fire against infantry and armored vehicles to defend the strong points that they have a mission to hold.

The withdrawal of outposts and artillery, possibly deployed in front of the position of resistance, is always carefully prepared, in time (time of withdrawal, synchronism to be ensured on the various parts of the front) and in space (axes of withdrawal, covered approaches). When the order is given, the movement is made by echelons under the protection of the artillery and infantry weapons of the position of resistance that are in a position to act effectively.

In particular, enemy armored vehicles can make it impossible, during the day, to withdraw from outposts.

Combat for the position of resistance.

424. As long as the attack is not imminent, the troops defending the position of resistance are stationed near their combat posts under conditions that allow them to carry out the prescribed work and spare them all useless fatigue.

When combat is engaged, the division commander leads it by maneuvering his fires and regulating the intervention of his reserves, with a view to maintaining the integrity of the position of resistance at all costs.

Operation of fires.

425. *When the attack is imminent,* the division general prescribes, according to the orders of the Army Corps commander, the execution of interdiction, counter-preparation, and possibly counterbattery fire. Frequently, the heavy weapons of the infantry will be able to collaborate usefully in the fires of interdiction and counter-preparation.

When the attack debouches, artillery and infantry try to break it by a common action; their tightly coordinated fire systems complement each other in the general barrage (No. 251) and in the main antitank barrage (No. 420); infantry barrage fire is particularly dense in zones that lend themselves less well to artillery action; stopping fires of the artillery more specifically interdicts the parts of the terrain that the infantry covers badly because of the flatness of their trajectories (No. 262).

The infantry concentrates its fire on the enemy infantry to force them into hiding and thus prevent them from exploiting the action of the tanks.

This phase of the fight will usually be decisive.

If the enemy gains a foothold in the position of resistance, the infantry and the artillery endeavor to limit its progress by stretching a continuous barrage of fire around the breach; they also use their fires to cut off any troops that break into the position from their reserves. The infantry stays concealed to the passage of the tanks to fire only on the infantry that follows them.

Use of reserves.

426. The division commander uses his reserves:

— to limit the local successes obtained by the enemy on the position of resistance;

— to restore the very integrity of this position.

In the first case, the reserves are used under the conditions prescribed in No. 269.

In the second, their mission is to counterattack and chase the enemy, who had previously been stopped, from parts of the position he had conquered.

The combat echelon units (regiments or battalions) execute immediate counterattacks with their reserved detachments; these detachments intervene without delay. Their action is supported by the fire of the automatic weapons and, if possible, by the elements of tanks being in immediate range as well as by the artillery able to intervene. Such actions must be studied and prepared for the various eventualities to be foreseen.

The *divisional* reserves can, after initially sealing off the breach (*colmatage*), carry out *more important counterattacks* based, like any attack, on the combined action of the arms, and consequently requiring a delay which one endeavors to keep to the minimum. Facing an enemy with armored vehicles, the use of tanks will be necessary. In anticipation of these operations, the division commander has reconnaissance carried out and the corresponding action plans drawn up. The need to act quickly will make it necessary to set the axis of the counterattacks according to the immediate possibilities of the artillery, to choose a restricted objective, and to use the reserves on the spot; the well-prepared action of artillery and tanks will always greatly facilitate their development and success.

ARTICLE 2.

WITHDRAWAL FROM ACTION.

427. In order to slow enemy progress by refusing close combat, the division commander will determine the composition of the first echelon according to the duration of resistance that he must offer, taking into account the orders of the Army Corps, the nature of the terrain, and the width of the action front.

The next echelon will be made up of the units available after having been placed, if necessary, behind the first echelon, at well-chosen points, the elements necessary for its withdrawal.

The different echelons are placed at a distance such that two successive echelons cannot be subject to the same deployment of enemy artillery.

In the withdrawal from action, it will generally be advantageous to place the regiments abreast in order to ensure the exercise of command in depth.

In order to ward off an attack by enemy weapons that are particularly to be feared in this phase of the fight, the obligation to cover each echelon with a natural obstacle on which the destruction will have been planned, will take precedence over any other consideration; it will not exclude besides the need to establish a system, even summary of antitank defense.

The withdrawal will generally have to be carried out at night unless certain favorable circumstances make it possible to attempt a disengagement during the day: for example, under the cover of fog, smoke, close terrain, or even if one has available armored vehicles, which are precisely suited to favor the breaking off of contact during the day to an infantry strongly pressed by the enemy.

It is often possible and always useful to deceive the enemy by leaving weak elements in contact with him to give the illusion of the effective presence of a unit already withdrawing.

Finally, an air defense system will cover, as far as possible, the withdrawal routes.

ARTICLE 3.

RETREAT.

428. When the division commander is ordered to conduct a retreat, he constitutes a rear guard with his available elements and establishes it on the line fixed by the Army Corps commander. He has it supported by a maximum of artillery and ensures its liaison with the rear guards of the neighboring divisions.

The engaged troops endeavor to hold until the night to carry out then their retrograde movement. If this is not possible, they fall back by day under the protection of detachments of the combat echelon held in place, then move out slowly and gain ground by disengaging the front of the rear guard.

The division commander directs his troops in the direction of the retreat assigned by the Army Corps commander. He indicates assembly points to units that must be handed over beforehand and specifies the conditions for carrying out the destruction that the Army Corps commander has ordered him to carry out.

The rear guard conducts a retreat in its turn when it has fulfilled its mission.

The considerations developed for the withdrawal from action with regard to the use of armored vehicles, antitank weapons, smoke, and defense against aircraft apply to the retreat.

ARTICLE 4.

SPECIAL CASES OF THE DEFENSE.

429. These kinds of operations are carried out in accordance with the rules prescribed in the corresponding articles of Title V (No. 277, 291, 292).

TITLE IX.

GENERAL POINTS ON THE USE OF LARGE CAVALRY UNITS.

CHAPTER ONE.

CHARACTERISTICS AND PRINCIPLES OF USE.

430. Large Cavalry Units are characterized by their mobility and firepower. Their offensive value is in direct proportion to their endowment in artillery and armored vehicles.

They are subject to constraints resulting from their size, their vulnerability, and their fragility, as well as the requirements of their maintenance and supply.

431. Depending on the mission received, the commander of a Large Cavalry Unit stages his maneuver by drawing inspiration from the guiding principles prescribed in No. 4. More than any other leader, he is called upon to make rapid, precise, and simple decisions with a spirit of enterprise in keeping with the mobility of his arm and the initiative required by the missions entrusted to him.

432. The maneuver of a Large Cavalry Unit must harmoniously combine the use of its different elements: horse-drawn, motorized, and armored.

It always tends to move quickly and securely, in spite of the terrain and in spite of the enemy, a powerful but voluminous mass of maneuver, with a view to concentrating the combined efforts of its assets on the most advantageous front.

It requires both centralized command and decentralized execution led by subordinates capable of initiative within the framework of their mission.

The commander of a Large Cavalry Unit mainly exercises his action and gives unity to the maneuver by using:

— *intelligence*, factor of his decision:

— *security*, which guarantees freedom of action;

— *the disposition,* which, by its deployment, makes it possible to carry out the maneuver and deal with the unexpected.

433. The combat of a Large Cavalry Unit involves, like that of other Large Units, the coordinated implementation of means and the rapid concentration of efforts on the front chosen for the main action.

In offensive operations, the commander of a Large Cavalry Unit first reconnoiters the enemy, tries to fix him on an extended front, and then attacks him in one direction with the maximum means. He does not hesitate to commit all his forces from the start, seeking success by surprise, power, and maneuver rather than by the succession of efforts.

To this end, he takes advantage of the mobility of his units to seize the adversary at the sensitive point by combining, if necessary, a front action and a wing action.

Against an adversary established on a defensive position, the Large Cavalry Unit fights like the Large Infantry Units; it must then be provided with the same means.

In defensive operations, their mobility and their high level of automatic weapons allow the Large Cavalry Units, supported on the terrain breaks, to resist on an extended front but for a limited time.

Their properties make them particularly suitable for delaying action. On the other hand, the defense without thought of retreat should only be asked of them in exceptional cases, and only with serious reinforcements in artillery and units capable of holding the ground.

CHAPTER II.

THE CAVALRY CORPS.

434. The Cavalry Corps can develop a rapid offensive or defensive action, the power of which varies with the nature[1] and the number of the divisions and elements of reinforcement that it encloses.

The Cavalry Corps commander assigns the subordinate Large Units their missions and defines the framework of their action according to his maneuver. He always keeps a reserve.

ARTICLE 1.

MANEUVER OF THE CAVALRY CORPS.

435. The Cavalry Crops commander, taking into account the considerations developed in No. 7, decides on his plan of maneuver.

---

[1] Cavalry Divisions—Light Mechanized Divisions—Motorized Infantry Divisions—Infantry Divisions, as appropriate.

He organizes, accordingly, the search for intelligence, security, and the marching disposition.

The *intelligence* that the Cavalry Corps command needs results from his maneuver intentions; he ensures the research by using aerial distant reconnaissance and ground distant reconnaissance, which are constituted by levies on divisional resources.

He sets the missions and directions of these two discoveries, he closely coordinates their action.

The Cavalry Corps commander determines, from the point of view of distant *security*, on which main lines of the terrain the divisions will place their security detachments in order to create around the Cavalry Corps the successive security zones within which he can move his maneuvering mass freely.

He specifies, from the point of view of close security, the missions and objectives of the advance guards of the divisions.

He organizes the air cover or addresses, in this respect, the necessary requests.

He distributes among the divisions, according to their mission, part of the additional resources allocated to him.

Finally, he decides on the *disposition* of the Cavalry Corps and the conditions for carrying out the maneuver.

To this end, he sets:

— the directions and zones of action of the divisions;

— the successive transversals that their main bodies must reach, the travel schedule or the guide unit, and the liaisons;

— the place of reserves;

— the measures to be taken to ensure control of the road network.

ARTICLE 2.

COMBAT OF THE CAVALRY CORPS.

436. The offensive combat of the Cavalry Corps generally implies a defensive attitude on the major part of the front and a concentration of the means on the direction of attack.

To this end, during the maneuver preceding the attack, the Cavalry Corps commander matures then adopts his decision, as and when intelligence on the enemy arrives; he chooses the direction in which he will concentrate his efforts and establishes his combat plan.

He then gives the divisions their missions, their directions, their objectives, and their zones of action and distributes, between them, part of the additional means that he has so far reserved or which it is possible to take in due time from Large Units that are holding passive parts of the front.

He keeps, in principle, under his direct orders, part of the artillery and always a reserve.

Certain situations (action on moving troops, pursuit) may justify several attacks, successive or simultaneous, separated by passive or contiguous fronts; all these attacks are equipped with means corresponding to their relative importance.

The Cavalry Corps commander coordinates, through the Cavalry Corps artillery commander, the actions of the divisional and Corps artillery.

437. When it comes to a withdrawal from action, the Cavalry Corps commander sets the successive positions that correspond to the different phases of the withdrawal, the conditions of their occupation in time, and especially the degree of resistance that he has decided to offer on each of them. He decides on the destruction plan.

The distribution of resources to be allocated to each position results in particular from this last decision.

When the Cavalry Corps must defend a position without thought of retreat, its commander defines the defensive position (position of resistance and outposts) and the missions of the divisions; he specifies the main directions to interdict and the points to be held. He complies, on the whole, with the general rules set out in this respect in Title V (No. 248).

In defensive operations, it is generally advisable to place the divisions abreast.

ARTICLE 3.

SPECIAL CASES OF USE OF THE CAVALRY CORPS.

438. *In covering*, forming a defense on a wide front, the Cavalry Corps commander applies the rules given in No. 292 on this subject.

He must keep himself in a position to withdraw from action if pressed by superior forces, and, for this purpose, he must maintain sufficiently numerous mobile reserves articulated on reconnoitered routes.

439. *On a battle front,* the Cavalry Corps can *exceptionally be* used as another Large Unit, taking into account its characteristics and, in particular, its ability to maneuver. It is then equipped with reinforcements corresponding to its current mission.

CHAPTER III.

THE CAVALRY DIVISION.

440. The horse-drawn and mechanized means of the Cavalry Division, its endowment with automatic weapons, and the motorization of its trains give this Large Unit power and mobility.

441. The Cavalry Division must receive a mission in relation to its means.

The orders given to him ***essentially*** include:

— a mission—a direction—a zone of action—objectives to be achieved under defined time conditions;

— the nature of the intelligence to be collected, the places and times to which it must be sent;

— what to do in the event of an encounter with the enemy.

The division commander establishes, consequently, his plan of maneuver from which result the successive orders by which he carries out his maneuver for the limited period which immediately concerns his subordinates.

These orders organize in particular:

— intelligence gathering (ground and aerial discoveries);

— security;

— the disposition of the division.

Discoveries and security are subject to prescriptions similar to those provided for the Cavalry Corps.

The disposition varies with the mission, the terrain, and the situation. In favorable terrain and if the necessary space is available, the motorized units, by their ability to establish contact quickly, find their use in the first echelon; the units on horseback follow on the road, as quickly as possible, oriented from afar towards the positions that the commander assigns them in his maneuver.

In difficult terrain and at a short distance from the enemy, it is, on the other hand, preferable to place mounted units supported by armored vehicles in the first echelon and keep motorized units in the second echelon.

The dual need to echelon the elements of the division in depth and to avoid mixing horse-drawn and mechanized units leads to the division of the Large Unit into tactical groupings adapted to the circumstances.

Each of the latter receives a mission, a direction, a zone of action, an indication of what to do in the event of an encounter with the enemy and successive bounds which allow the commander to coordinate his maneuver.

To avoid the dispersion of efforts and reserve the means to exert his personal action on the combat, the commander of the Cavalry Division limits the means that the grouping commanders can engage on their own initiative.

442. *The offensive combat* of the Cavalry Division engages and develops according to the rules prescribed in chapters I and II of this title.

The division general chooses the main point of effort, either on the enemy's front or on one of his wings, depending on the purpose of the attack.

The action is supported by all the artillery and armored vehicles.

443. *The defensive combat* of the Cavalry Division can develop on an extended front if it relies on the terrain breaks; it implies the constitution of very mobile reserves (motorized dragons in all-terrain vehicles, motorcyclists, armored cars, and tanks) allowing maneuver. Faced with a powerful adversary, the duration of resistance is limited by the weakness of manpower and artillery.

In the withdrawal from action the division general establishes his plan so as to defend, with part of his means, a first position and to prepare with other elements the defense of a position located further back, at a distance of the first sufficient to oblige the enemy to provide two successive efforts.

The choice of positions is dominated by the interest presented, for the maneuver, by terrain breaks and obstacles. The division commander sets, according to the total time that he has been prescribed to gain the ground and taking into account the particular defensive value of the successive positions, the duration of the resistance that he intends to offer on each of them.

The division general also indicates to his units the organization of the command, the successive positions, the disposition, the distribution of the elements between the different positions, the role of the artillery, the conditions of the withdrawals. He complies with the instructions of the higher authority regarding the destruction to be carried out.

In the *defense without thought of retreat* the Cavalry Division fights like an Infantry Division. It is then equipped with the necessary reinforcements (artillery, machine gun battalions, etc.). The division commander takes care of the deployment and protection of the encampment of horses and carriages, the safeguard of which is the essential condition for the continuation of subsequent operations.

In the withdrawal, the Cavalry Division is protected by rear guards reinforced with armored vehicles. The elements withdrawn from combat quickly fall back to the area fixed for the resumption of a new maneuver disposition.

The Cavalry Division can be employed *in cover* or *on a battle front* under the conditions prescribed concerning the Cavalry Corps.

CHAPTER IV.

THE LIGHT MECHANIZED DIVISION.

444. The Light Mechanized Division is a force characterized by its mobility, by its radius of action, and by the protection enjoyed by the firepower weapons.

Equipped for the distant search for intelligence, capable of ensuring its own security, it can fulfill, with the necessary reinforcements, all the missions devolved to the Large Cavalry Units; it can, in particular, provide the intelligence and security essential to motorized Large Units.

Its numerous armored vehicles allow it to provide brutal and sudden actions *in the offensive*, while its endowment with automatic weapons puts it in a position *in the defensive* to effectively interdict the terrain on extended fronts.

The proportion between the motorized elements and the armored elements of the Light Mechanized Division determines their use.

The motorized elements, by their ability to occupy the terrain, can quickly establish curtains of fire, support the attack of armored vehicles, and occupy the objectives conquered by them.

The armored formations constitute the dynamic element of the Light Mechanized Division. Their modes of action are attack and counterattack, but these must be protected against the adversary's antitank weapons by the fire of automatic weapons and antitank dispositions, by artillery, and by aviation; their success is made definitive only by the occupation of the conquered ground by the mounted elements alone capable of ensuring it.

More than any other formation, the Light Mechanized Division, to develop its offensive or defensive properties, must be able to absorb and use the reinforcement units appropriate to the particular mission assigned to it.

These reinforcements will include, in particular:

— in defensive operations, antitank formations, artillery and infantry units, particularly machine gun battalions, etc.;

— in offensive operations, aviation, tank units, artillery units, preferably towed, possibly motorized infantry, etc.

The Light Mechanized Division must have, for the implementation of such reinforcements, means of command and, particularly, substantial signal communications.

445. The main missions that can be entrusted to the Light Mechanized Division are:

— *a reconnaissance mission for* large units; the Light Mechanized Division is capable of launching its distant reconnaissance detachments on a front of 40 to 60 kilometers at a distance which can reach 100 to 150 kilometers at a speed of advance of 15 to 20 kilometers per hour;

— *a security mission:* at the beginning of operations, in cover or in reserve cover, during operations, whether to cover the entry into action of motorized forces or to protect the front or the flank from a distance forces carrying out an offensive or defensive maneuver;

— *a battle intervention mission:*

In the *offensive* battle, it can, after a penetration in the enemy's front, burst through the communication route network, or quickly overrun one of the opposing wings by linking its action in time with the attacks of other Large Units. After a success, its qualities of speed find their fullest use in the pursuit.

In the *defensive* battle, it can perform a withdrawal from action or, held in reserve, intervene to plug a breach or to counterattack a victorious enemy.

The combination of the action of the air forces and the Light Mechanized Division is particularly fruitful.

446. The principles of use of the Light Mechanized Division are the same as those prescribed for the Cavalry Division; it is only necessary to apply them in a broader framework, in a shorter timescale and, for this purpose, to take particularly quick and simple decisions.

In offensive operations, the commander of the Light Mechanized Division organizes his distant reconnaissance and his security taking into account the need to be informed all the more quickly and protected all the further as his mass of maneuver is bulkier and faster.

He determines the marching disposition of the Light Mechanized Division in such a way that it can be transformed without delay into a combat formation; he ensures the mastery of the road network necessary for its maneuver.

The Light Mechanized Division is articulated into tactical groupings like the Cavalry Division. The constitution of these groupings often requires the *temporary* joining of units borrowed from the various elements of the division under the orders of the same leader.

The offensive maneuver of the Light Mechanized Division is characterized by its rapidity. To this end, the leader applies the rules prescribed in No. 432.

In principle, the attack starts from a base of departure, made up of the motorized elements, their debouchment supported by their own fires and those of the artillery. In the presence of a surprised enemy, it is often advantageous to debouch brutally with armored vehicles, with no other support than that of their own fire.

The commander of the Light Mechanized Division always keeps a reserve; he provides for the occupation of a fallback position in the event of failure.

In defensive situations, the Light Mechanized Division is able to quickly establish a curtain of fire, carry out effective stopping actions, participate in a withdrawal from action, and in particular, cover the withdrawal of an echelon to its subsequent position

TITLE X.

GENERAL INFORMATION ON THE USE OF MOTORIZED LARGE UNITS.

SINGLE CHAPTER.

ARTICLE 1.

CHARACTERISTICS.

447. Motorized Large Units include:

— organic corps elements with full motorization;

— motorized Infantry Divisions whose units organically possess:

— some, the means of transporting their entire strength;

— the others, the means of transporting only part of their strength.

From their composition, it follows that the motorized Large Units must make extensive use, for their integral transport, of extra-organic formations.

448. Strategic mobility is the essential characteristic of these Large Units.

Their tactical employment benefits from their ability to *rapidly carry out* reconnaissance, establish the liaisons essential to combat, and move their artillery and reserves, as well as to carry out supply operations, even from distant bases.

Motorization imposes on them, on the other hand, rather heavy constraints. Linked to the roads for most of their elements, bulky, and vulnerable, ***motorized Large Units are incapable of ensuring their own security, although this is the imperative condition for the movement and unloading of their main bodies.*** During transport, all action is forbidden to the latter; they must unload before entering the battlefield.

449. From all of their possibilities, it follows that the motorized Large Units:

— form par excellence, in the hands of the high command, *reserves* endowed with great strategic mobility;

— require their *command,* at all levels, to adapt to the particularities of motorized maneuvers;

— *can only move their main body under the shelter of a cover* which absolutely guarantees their security on their front and on their flanks;

— are used after unloading, *in battle or in combat under the same* tactical conditions as non-motorized Large Units.

ARTICLE 2.

MISSIONS AND CONDITION OF USE.

450. Motorized Large Units are particularly suitable for carrying out missions that offer a tactical or strategic priority, namely:

— *in battle,* to carry out the rapid concentration of reserves at the desired point, either with a view to the development or completion of an offensive action, for example, for a decisive thrust, the extension of the front, or the pursuit of disorganized troops, or with a view to recovery after a failure (closure of a breach, counterattack);

— *in open terrain,* to reach and use before the enemy a region favorable to the development of the maneuver, for example, to outflank or envelop a wing, carry out a covering, seize opportunities, or act against a line of communication.

451. The conditions of movement differ *according to whether or not it takes place under the shelter of a pre-existing front:*

— in the first case, the motorized Large Units can proceed, in security, to their movement and their unloading under the ordinary conditions of transport of troops by motorized formations;

— in the second, on the other hand, the use of motorized Large Units can only be envisaged in cooperation with elements capable of providing for the safe movement, encampment or unloading of their main bodies by establishing effective cover forward and on the flanks of their zone of movement.

It is advantageously constituted, for this purpose, a *motorized grouping*, bringing together, under the same command:

— Motorized Large Units;

— Cavalry Divisions (Light Mechanized Divisions or Cavalry Divisions) and air forces, in a sufficient number to guarantee the security required by the conditions of the general situation at the time.

The motorized grouping receives, in addition, the motorized train groupings necessary for its transport as well as reinforcements appropriate to its mission (motorized artillery, motorized machine gun battalions, units of tanks and antitank weapons, etc.).

ARTICLE 3.

COMMAND.

452. The maneuver of a motorized grouping is conducted by taking inspiration from the permanent guiding principles (No. 4), but it is influenced by the frequent distortion of situations, by the impossibility for the commander to make his plan far from an enemy often still not fixed, and by the speed with which the command will have, when the time comes, to make its decision without provoking either false operation, or extended halt of the columns.

These peculiarities of motorized maneuvering lead the leader to:

— obtain intelligence quickly and distant, both on the enemy's activity, means and conditions of movement, and on the characteristics of the zone of approach;

— demonstrate speed and simplicity in decisions relating to the movement of units or columns, the choice of their unloading zone, and the deployment that will follow.

For this purpose, the commander of a motorized grouping must largely precede his main bodies, carefully organize his movements, and divide his headquarters by using, in particular, an advanced command post, reduced, mobile, and chosen according to the *existing and immediately developed signal communications network.* His successive orders will be brief, and the liaison and signal communications system adapted to these command procedures.

The leader must be assisted by a staff capable of expressing his decisions in clear and concise orders and by subordinates capable of planning and initiative within the limits of the missions entrusted to them.

ARTICLE 4.

RECONNAISSANCE AND SECURITY.

453. When a motorized grouping is called upon to move *on open terrain*, without benefiting, consequently, from the cover of a pre-existing front, *the reconnaissance* is ensured by the air forces and by Large Cavalry Units operating in concert, *security* is guaranteed by the combined action of Large Cavalry Units, reconnaissance battalions, advance guards and flank guards.

454. Reconnaissance is organized according to the ordinary rules (No. 128).

455. The security of the movement, encampment or unloading of motorized groupings depends on the occupation ***before the movement of the main body*** of the terrain breaks or lines favorable to the defense that cover the encampment or unloading zone ***that the grouping must reach at the end of a day's march*** and those which interdict access to the flanks of the zone of march.

The security units must therefore be advanced, and then established:

— forward, at distances corresponding to the length of the stage that the leader has decided to have his main body cover, taking into account the situation, the speed of advance of the grouping and that of the adversary, estimated from the intelligence collected on the nature of the means of transport used by the latter;

— on the flanks, at suitable distances.

It follows from these considerations that the security of a motorized grouping requires different methods from those set out in Title III (No. 134). In this particular case, all security effort has to be exercised for the sole benefit of covering the arrival staging area and open flanks.

With this in mind, the advance guards and the flank guards no longer receive the mission of guaranteeing close security but of participating in distant security:

— or, if the terrain is favorable and if the situation so requires, by supporting the defense already organized by the Large Cavalry Units;

— or by giving depth to this cover and installing itself, for this purpose, on an advantageous position located behind and close to that held by the Large Cavalry Units.

The advance guards and advance flank guards are informed and covered by the reconnaissance battalions of the motorized Large Units and conform their maneuver to the intelligence they receive from the Large Cavalry Units.

456. The commander of the grouping specifies the mission of the various security bodies.

He reinforces, if necessary, the reconnaissance battalions with motorized infantry, armored vehicles, and antitank weapons.

He gives the advance guards and the flank guards a composition in relation to the defense effort they will have to deploy, taking into account the particularities of the terrain.

He takes care to provide them with the troops of engineers and the matériel necessary to clear the obstructed routes and to restore the passages in the event of destruction conducted by the enemy.

He is informed by the cavalry on the state of the road network and attaches a technical staff to it for this purpose.

He pushes the elements intended for the organization of the traffic in the wake of the security bodies.

457. *Air cover* is ensured under the general conditions prescribed in Nos. 301 and following. The commander of the grouping organizes, for this purpose, a command of the air forces and antiaircraft defense.

Each unit also takes special care of *its local security* (No. 150 and 156).

ARTICLE 5.

MOVEMENT.

458. When the security units are established in their positions, the movement of the main body can take place inside the security polygon thus created under conditions which, excluding the possibility of an encounter with the enemy, make it possible to obtain the maximum efficiency from the motorized transport.

459. To this end, the movement of the main body is organized, broadly, in such a way:

— to make full use of the road networks in order to reduce congestion and the vulnerability of the march disposition;

— to make this disposition as flexible and less vulnerable as possible by articulating its various columns into ordered groupings;

— to lighten the combat units by pushing back all the elements of the services which are not essential for them to engage;

— to safeguard the order and precision of movements by meticulous preparation and organization of traffic, by exact technical discipline and by the proper functioning of liaison and signal communications.

460. The columns are deployed either by combined arms tactical groupings or by march groupings that include detachments of the same arm or same mobility. The first process is essential for security dispositions intended to engage as soon as their motorized elements have been unloaded; the second generally applies to Large Units or main body units.

461. The motorized grouping presents, in its order of march, a surface and, therefore, visibility and a vulnerability such that the approaches of the main body can only take place during the day under the protection of vigilant air defense and by endeavoring to reduce the risks by echelonment the columns in time and space. *Night movement,* despite its reduced performance, will most often be prescribed for the main body. It will always be advantageous to carry out the entire movement of the main body during the same night. This will thus avoid compromising the secrecy of the movement by starting or prolonging it during the day.

ARTICLE 6.

DEPLOYMENT AND BATTLE.

462. Despite strong air and ground cover, the *deployment* of a motorized grouping is the most delicate part of its approach march.

The unloading of the main body of the transported units should, as a general rule, ***only take place outside the range of the adversary's long-range heavy artillery fire*** and, *as far as possible*, under cover of darkness or, failing that, terrain covers.

After unloading, the main body of the motorized Large Units are routed to their attack or defense positions according to the rules governing the approach marches and placement of other Large Units.

463. The conduct of the battle is ensured by the commanders of the motorized Large Units under the conditions previously prescribed for the non-motorized Large Units.

TITLE XI.

SPECIAL CASES OF USE OF LARGE UNITS.

SINGLE CHAPTER.

ARTICLE 1.

MOUNTAIN OPERATIONS.

464. Large Unit operations in the mountains are not dealt with in this Instruction. They are the subject of an attached Instruction.

ARTICLE 2.

RIVERS.

465. Rivers constitute, after the destruction of their crossing points, one of the most important obstacles that can oppose the advance of a Large Unit or, on the other hand, reinforce the value of its defense.

I. Crossing a river.

466. While it is relatively easy to get light elements of cavalry and infantry to cross a river, the crossing of the main body of the Large Units requires the prior implementation of the more considerable means when the river is wider, deeper, and faster. Gathering the necessary means must, therefore, be the subject of long-term forecasts because of the time required.

In most cases, it is important that the units arriving first on the obstacle can immediately undertake its passage by *surprise,* without giving the enemy time to organize or reinforce its resistance. It is, therefore up to the commander of a Large Unit to take all measures so that these leading units approach the river in the disposition and in the zones that he considers most favorable for crossing it; he distributes in advance, between the units whose success he deems most useful to his maneuver, the means of organic passage and reinforcement as well as the specialized engineer personnel at his disposal.

A surprise crossing is attempted on the entire front by all units that reach the river; these units first try to throw small elements onto the enemy bank by swimming or using intermittent organic or makeshift means of passage.

Leaders of all ranks have a duty to exploit any local success without delay by creating small bridgeheads on the enemy bank, of a breadth and depth commensurate with the numbers at their disposal and covering the initial crossing point.

These bridgeheads are gradually reinforced and enlarged by frontal and lateral actions.

Amphibious armored vehicles can cross the deepest rivers on their own, but their use requires prior reconnaissance and, in some cases, access works. If the vehicles are propelled on the surface by using a propeller, the reconnaissance must focus on the points where the banks allow them to be launched and landed; if the vehicles travel on the bottom of the water on tracks, they must also relate to the nature of the bottom of the river and most often require soundings.

467. When faced with the presence of an enemy established on the opposite bank whose fires have caused the failure of the surprise crossing attempts, a *forced* crossing is carried out, after combining the means. It is undertaken on several zones large enough to avoid the concentration of enemy fires. These zones are chosen by the leader depending on his subsequent maneuver, taking into account the technical possibilities.

The operation consists of three phases:

— *the crossing of the first elements* (cavalry or infantry, then small units of armored vehicles and detachments of light artillery) by temporary means and in certain cases by footbridges;

— *the construction or restoration of bridges;*

— *the crossing of the main bodies.*

468. The operation is carried out under the protection of artillery, cavalry or infantry fire installed on the bank in such a way as to effectively cover the opposite bank and constitute a bridgehead of projectiles, under the shelter of which the cavalry or infantry elements cross the river on intermittent means of passage and possibly on footbridges.

These first elements are heavily equipped with antitank weapons.

Amphibious armored vehicles or vehicles transported on intermittent means can in certain cases cooperate effectively in the enlargement of the first bridgehead thus created.

Some elements of light artillery can then switch to intermittent means.

The troops which cross a river first must be chosen from among those who have occupied the friendly bank for some time and are thus accustomed to observing the river and the enemy bank.

469. Only when the bridgehead is deep enough and strong enough to shield the river and its banks from the aimed fire of the automatic weapons of enemy infantry and the sights of nearby enemy observation posts can the construction of the bridges be undertaken with chances of success. The order is given by the command.

During the works, the movements by intermittent means or on footbridges continue, in the axis of the engaged units, to feed the combat.

The attack, in fact, continues without stopping, with a view to enlarging the bridgehead and thus putting the crossing points more and more completely sheltered from the fire of the enemy artillery.

The defense of crossing points against air attacks and incursions by armored vehicles must be one of the constant concerns of the command.

470. The passage of heavy elements from main bodies begins as soon as the bridges are open to traffic. It is regulated by the command.

The order of the passages is established according to the development of the operations.

It is necessary during this phase to avoid initially accumulating too many elements on the enemy shore, as long as the depth of the conquered terrain is not sufficient to approach them in complete security.

The number of bridges to be established for each Large Unit must be sufficient to allow not only movements and supplies forwards, but also large evacuations to the rear.

471. Night and fog are advantageously used for river crossings, specifically for the crossing of the first elements.

A particularly favorable time is shortly before daybreak, as it allows the combatants to conceal their crossing in the last shadows of the night and the very frequent fog at this time while seeing clear enough to move when they approach the enemy bank.

When the crossing is conducted by force, the artillery takes all necessary measures so that its first fires, triggered at zero hour, are immediately effective.

Artificial smoke sheets and screens provide very effective protection for all phases of a crossing operation.

Their use, like the use of night or fog, makes it possible to advance the moment when one can begin to build the bridges.

II. Defending a river.

472. Defense without thought of retreat of a river can require a high density of occupation because of the woods, bushes, mills, factories, etc., that often cover the banks.

The fire system must effectively cover the enemy bank, the actual course of the river, nearby cover likely to shelter personnel and matériel, as well as the access routes to the enemy bank.

The general barrage will most often be placed on the river itself, the main line of resistance being realized by the very edge of the friendly bank.

The effort of the defense will relate to the points that seem most favorable to an enemy attack, from both a tactical and a technical point of view.

The antitank defense will be reinforced on the parts of the river that best lend themselves to attempts by armored amphibious vehicles.

Some pieces of light artillery can, in certain cases, be placed in caponiers on the friendly bank.

The defense will increase the value of the obstacle by raising the level of the water using dams, widening the banks on both banks, putting the fords out of service, immersing chevaux-de-frise and dormant mines, or laying iron wires on the very outskirts of the friendly shore.

The position on this bank must have sufficient depth and allow the establishment of switch trenches intended to check the expansion of any bridgehead created by the enemy.

The antitank defense is also organized in depth and on switch trenches to control the action of armored vehicles that have succeeded in forcing the obstacle.

If the enemy manages to maintain a bridgehead, the artillery, and aviation, while cooperating in stopping his advanced elements, take as their main objectives his crossing points, which can also be attacked by drifting mines launched, preferably during the night, upstream of the front attacked by the enemy.

ARTICLE 3.

WOODS AND VILLAGES.

473. Forests and agglomerations make it possible to hide movements (in particular: approaches, setting up, breaking out of combat) and encampments from ground and aerial view. To a certain extent, they protect against enemy fire and allow the inexpensive creation of shelters and obstacles.

On the other hand, they break the cohesion of large and small units that enter or occupy them; they impose serious difficulties on the use of the various weapons and, more especially, on their combination. They are particularly unfavorable to a strong action of armored vehicles, as much by the scarcity of practicable zones as by the ease of creating obstacles there.

474. *In defensive operations,* occupying regions of close terrain initially requires installing numerous infantry. Thereafter, defensive works make it possible to create impassable passive zones, and manpower can be reduced.

Woods and settlements are natural strong points that provide the defender with the ability to establish weapons away from armored vehicle attacks and from ground and aerial sight. They are occupied in first priority; later, the progress of the works will make it possible to extend the defense outside them. The defense effort must bear as much on the flanks and the rear as on the front.

Woods and settlements are prime targets for enemy artillery and air force projectiles; when their dimensions are small, they can become veritable bomb nests.

The main line of resistance is generally not installed at the very edge, often exposed to precise fire. Inside, successive lines of fire are established, and redoubts are organized to control the outlets towards the rear. Finally, a reserve (infantry and armored vehicles) is maintained near the redoubts or in a strong point located further back, to oppose outflanking maneuvers.

In offensive operations, large forest regions and agglomerations are to be avoided for rapid and decisive actions.

Extensive woods and agglomerations are not, in principle, the object of deep attacks; their outflanking is the rule. Manpower of varying strength is assigned to the attack on their edges, while the main body of the forces carries out the outflanking maneuver. Containment followed by mopping up may be necessary.

However, due to their size or location, direct attacks may be required.

After conquering the front edge, the attack strives to advance by the side edges; if it does not succeed, the terrain is conquered in successive stages; the next section is attacked only after the previous section has been cleared and the attacking troops have been put back in order.

The conquest of each successive section is protected by large box barrages, the fires of which are intensified on the clearings, open spaces, and crossroads.

Armored vehicles cooperate advantageously in the conquest of the edges, then working alone or in small units, accompanying the elementary infantry units, they can provide a powerful aid to their advance along the paths, roads, or forest paths that check their direction.

Behind the echelon of fire, well-in-hand reserves occupy the lateral roads and the crossroads to ensure possession of the terrain and check the panic always to be feared in such a case.

The main body of artillery formations is never engaged in forests or settlements until the opposite edges are in possession of the attacker.

475. Whatever the phase of combat envisaged, *keeping the units in good order on their direction* is an essential condition for success.

Reorganization of the disposition takes place all the more frequently as the course of the terrain is more difficult; lateral roads are particularly suitable for these operations.

Combat in the woods or in villages requires troops who are particularly disciplined and trained in this kind of combat, during which the soldier always remains very impressionable. All means must be implemented to maintain the cohesion of the units.

ARTICLE 4.

NIGHT OPERATIONS.

476. The activity of the air forces and, even more, that of armored vehicles is greatly reduced during the night. Without observation, the fire of ground weapons assumes a systematic form that reduces its effectiveness.

Night promotes surprise.

The night is advantageously used by the leader to conceal the movements of his unit.

Night movements, however, impose serious fatigue on the troops. March speed is noticeably reduced for foot and horse-drawn units; it is reduced to one-third for motorized units operating with all lights off.

Close to the enemy, night will be frequently used during the end of the approach march and the setting up of units for the attack:

— to cross areas of terrain, mainly ridges, seen or beaten by the enemy, to put the attacking infantry and armored vehicles in place, to set up bases of fire and advanced observation posts, to deploy the artillery, and finally to carry out, if necessary, a relief of troops in contact or a passage of echelons;

— to attempt to cross a river by surprise;

— to break off contact, by withdrawing the main body of a unit to the rear during the night, covered by light first-line elements left in place until daylight.

The movements required by these operations are generally carried out across fields and are very slow. Their proper execution requires daytime reconnaissance and meticulous preparation.

477. *In the offense*, night operations most often aim either to complete the conquest of an objective or to seize part of the terrain necessary for the debouchment of an attack.

Launched in the first part of the night, they make it possible to use the second part for organizing the conquered terrain. On the other hand, their launching in the second part of the night can make it possible to avoid an always dangerous nighttime enemy reaction.

A night attack is prepared in every detail, but it cannot be controlled.

It requires officers, noncommissioned officers, and troops to have an exact knowledge of the terrain, directions, and precise objectives that are clearly recognizable on the ground. The artillery and the heavy infantry weapons support and protect these operations by the same methods of fire during the day. However, the fires are more systematic; they are executed according to a schedule and using very simple signals. Safety distances are increased.

The first echelon of infantry does not fire, it marches to contact with the enemy.

These operations are generally detailed, carried out in limited numbers on narrow fronts, and with nearby objectives.

Nevertheless, in the presence of an adversary fire system without depth, or in a very compartmentalized terrain, night attacks can be executed on an extended front. In this case, it is advisable to organize distinct attacks on sufficiently spaced directions so that the reactions of the enemy on one of them do not have an effect on the others.

Coherence of efforts is achieved by selecting targets and determining the time of attack.

Although there is no maneuver during combat at night, it may be useful to maintain a reserve behind the attack so as to ward off a possible incident; it will most often act as a fallback echelon.

478. *On the defensive,* the units fight against the difficulties of night combat through the careful registration of all weapons and a well-organized lookout service, as well as by the use of flares.

By the systematic fire from the defensive weapons, the command strives to frustrate the adversary's night actions.

The defensive system at night may differ from that of the day. Thus, the defense of an interval between two strong points, ensured during the day by crossfire, may require effective occupation by night.

Their knowledge of the terrain can allow small units to successfully execute immediate counterattacks.

Patrols or raids can be conducted at night to seek specific intelligence and, in particular, take prisoners. These operations require extensive preparation, regardless of their importance.

During nighttime, security is ensured by special procedures. Distant from the enemy, guarding access roads may suffice, while on the other hand, near the enemy, the lookout service must be reinforced and extended, especially around the directions which the enemy may follow.

Lookout posts often occupy different locations from daytime ones to avoid being abducted and also because, at night, observation gives way to listening.

479. *In the offensive, as in the defensive,* the night can be used to seek, verify, and maintain contact.

When the command deems it useful to carry out such operations, it instructs the artillery to provide momentarily in front of the first echelon of infantry a zone free of fire and sufficient to allow the search for contact.

It sets a very clear line of the terrain up to which the first line infantry units are authorized, in the event of the enemy's withdrawal, to push at night and without new orders from the occupying elements, without giving up the terrain which previously held and whose occupation, generally maintained until daybreak, never ceases without a specific order.

The night is finally put to good use for supplies and evacuations, as well as for the execution of certain works on the parts of the terrain seen or beaten by the enemy. Harassment and interdiction fire by aviation, artillery, and heavy infantry weapons can frustrate and even prevent these operations.

Whatever the night operation envisaged, the commander must take the greatest account of the atmospheric circumstances and the lunar conditions of the moment. It is important, in fact, to give surprise its full value, not only to conceal any movement from enemy observation or listening, but also to put the troops in a position to camouflage their new installation before daybreak.

Operations in foggy weather are conducted according to the rules given for night operations; all precautions are taken in case the fog suddenly dissipates.

Emissions of *artificial smoke,* in the form of sheets or screens, give an operation most of the advantages of night, while retaining some of the facilities afforded by day.

It is essential to deploy them on a broad front, or simply on several points of the battlefield, to prevent the enemy from locating the operation you are trying to conceal.

TITLE XII.

OPERATION OF THE SERVICES.

CHAPTER ONE.

GENERAL PRINCIPLES.

480. The regularity of the functioning of services is one of the essential conditions for the smooth running of operations; it has a considerable influence on the morale of the troops.

The command gives orders to the service chiefs; he sets their missions, indicates to them the conditions under which these missions must be fulfilled, and provides them, if necessary, with additional personnel, matériel, and means of transport; the distribution of these resources is based not only on the requests of the service chiefs but also on the degree of priority that he attributes to the various needs.

The command must consider the possibilities of the various services in the operations it prepares; therefore, it is essential that he be constantly aware of their situation.

To this end, the command at each echelon has his service chiefs with him.

Within the limits he deems compatible with the secrecy of operations, he keeps them informed of his intentions, so as to enable them to make the necessary forecasts.

Command decisions are brought to the attention of the services, either by the second part of the general operations orders, or by specific orders or instructions.

In each Large Unit headquarters, a bureau is generally responsible for drafting and transmitting the relevant orders and directives to the services; it monitors their execution.

The officers who make up this bureau must have a perfect knowledge of the functioning of the services; they maintain close contact with them and are constantly in a position to inform the command about their situation. Some of them belong to services (intendancy, health).

Chiefs of service are personally responsible to the command for the execution of orders relating to the employment of their service.

Within the framework of these orders, any service chief, at each echelon, has full and complete initiative to ensure the execution of his service; he gives the chiefs of services at lower echelons the technical instructions he deems necessary, keeps abreast of the operation of their services, and helps them to the extent compatible with his own resources.

CHAPTER II.

SUPPLY AND MAINTENANCE SERVICES.

ARTICLE 1.

GENERAL.

481. Supply and maintenance services are generally represented in all Large Units by chiefs of service with specialized formations or bodies and various supplies (food, ammunition, matériel, etc.).

The service chiefs meet the needs of the troops with the help of supplies that they have themselves or those that they cause to be sent by the service chief of the higher echelon.

Provisions are transported on vehicles following the troops (mobile reserves) or stored in warehouses or depots echeloned towards the rear and preferably placed near the railroad lines (fixed reserves).

Mobile reserves allow service chiefs to meet daily or urgent needs.

Fixed reserves are intended for:

— to replace the supplies of the mobile reserves which would have been consumed;

— to meet certain normal needs of the troops by daily shipments;

— to meet the exceptional needs of the troops.

The role of mobile reserves is essential; alone, they make it possible to deal with the contingencies that arise in mobile warfare. When circumstances cause a Large Unit to remain temporarily in place, the command determines the relative importance of the mobile reserves and the fixed reserves.

ARTICLE 2.

ARTILLERY SERVICE.

Supply of ammunition and gasoline.

482. Artillery ammunition supply operates at each echelon under the direction of the artillery commander for that echelon.

The artillery commander of the army, guided by the directives and projects of the command:

— draws up ammunition requests;

— proposes the locations of army depots;

— ensures the distribution and, if necessary, the storage of ammunition allocated by the higher authority.

For these various operations, he has the director of the Army Ammunition Service, who himself has under his authority the Army Ammunition Park.

In each Army Corps, the artillery commander distributes, according to the directives that have been given to him by the command, the ammunition allocated to the Army Corps between the divisional artillery and the artillery battalions directly at his disposition.

For these operations, he has the *artillery park of the army corps* and, possibly, *divisional parks;* he can exceptionally call upon the vehicles of the artillery units, organic or not, placed under his command.

If these means are insufficient, he asks for the assistance of the transport units of the Army Corps, then of the Army.

The Army assists the Army Corps less by providing additional means of transport than by pushing the ammunition it allocates to them as far forward as possible.

The divisional artillery commander plays in the Infantry Division, for the supply of ammunition, a role analogous to that of the commander of the corps artillery in the army corps.

Infantry and aeronautical ammunition, grenades, bombs, pyrotechnics, and explosives are generally stored at each echelon in the same depots as artillery ammunition. The reserve air forces, in principle, use specific depots for their ammunition.

The distribution and transport of ammunition are the subject of orders given directly by the command and executed, each as far as it is concerned, by the chiefs of the artillery service and the officers in charge of transport.

483. The supply of gasoline and lubricants to Large Units and aeronautical units is ensured, in the Army, by a *gasoline and lubricant park* which operates under the orders of the director of the Army's ammunition service.

The Army receives gasoline either in fixed installations (*main gasoline depots* supplied by the general headquarters and managed by the Army gasoline park) or directly at gasoline depots located on a railroad or river. For the transport of gasoline, the Army has tank truck companies that transport the collected gasoline to *secondary depots* where it is packed or that deliver it directly, in delivery centers, to units equipped with tank trucks.

Replacement and repair of matériel.

484. In an Army, replacement and repair of rolling matériel and armaments is carried out under the direction of the artillery commander, and the director of technical services, who has specialized *repair parks* (*parcs de réparation*) for each category of matériel—artillery carriages, military train, small arms—for this purpose.

These parks can be broken down into two parts:

— *a light echelon,* capable of being moved easily and likely to be set up close to the combat front;

— *a heavy echelon.*

Matériel that these parks cannot repair are sent to the *depot parks* (*parcs entrepôts*), which are also responsible for sending replacement matériel to the army.

In the Army Corps and in the Division, matériel replacement and repair are carried out by the *Army Corps repair team* (*équipe de réparation de corps d'armée*) and the *divisional repair team* (*équipe de réparation divisionnaire*), respectively, which are part of the Army Corps and divisional artillery parks.

In principle, these units only carry out minor repairs that are likely to be carried out quickly.

485. Army *tank parks* supply tank units with all types of matériel. They carry out repairs they can perform and direct the matériel they cannot repair to the interior.

They are placed under the orders of army tank commanders.

ARTICLE 3.

ENGINEERING SERVICE AND WORKS.

486. The various engineering services function, in an Army, under the high authority of the commanding general of army engineering, who has:

— an engineering services directorate;

— an army engineer park.

There is also a *works directorate*, which is placed under the same authority as the engineering services directorate.

I. Engineering Services.

487. The director of an army's engineering services is in charge of everything concerning the supply of tools and matériel to the troops, the maintenance of the army's road network, the camps and cantonments, the water service, general electrical power needs, and forest exploitation.

Matériel Service.

488. The operation of the Matériel Service is based at all echelons on the constitution of depots, fixed or mobile, echeloned in-depth and maintained both by exploiting local resources and the aid of shipments made by the higher echelons.

Roads Service.

489. At the army echelon, the Road Service is headed by a field-grade officer, the Chief of the Road Service. He ensures, in principle, the rehabilitation of existing roads, the creation of new roads, and the exploitation of quarries in the Combat Zone.

However, part of his personnel and matériel may be temporarily placed at the disposal of a subordinate Large Unit to be employed on the roads that this Large Unit is responsible for maintaining.

Conversely, the Large Units and the services of the army provide engineers to the army, for the execution of road works, with the personnel and matériel that the command deems useful to allocate to it.

The engineer commanders of army corps and divisions use in the execution of road maintenance works, in the zone for which they are responsible:

— engineer units placed under their command;

— units of laborers placed temporarily at their disposal by the commander of the Large Unit to which they belong;

— possibly, units of specialized laborers, taken from the road service of the army.

In each army, the matériel necessary for road works is managed by the chief of the army road service, who has special repair workshops.

Camps and Cantonments Service.

490. The Camps and Cantonments Service is responsible for improving the conditions of habitability of the camps and cantonments where the troops are called upon to pass through or stay. For this purpose, on the territory of each army (staging zone), a certain number of zones is planned. Each of them corresponds, in principle, to the encampment of a division. Service in the camps and cantonments is provided by a zone

Major, who is placed under the direct authority of the commander of the unit occupying the zone and receives only technical instructions from the camp and cantonment service chief.

Water Service.

491. The Water Service provides for the needs of the troops by using existing water points and wells. It creates new ones as needed or carries water to supply centers; transport is carried out by using special vehicles (water barrels and horse-drawn or motorized tank cars, tank wagons).

The problem of supplying the troops with water becomes especially important during a period of stabilization and when the operations in progress lead the command to unite numerous forces in a limited space.

The water service also considers all geological studies and research that may be necessary for its proper functioning or even useful to the command and other services.

Electrical Service.

492. The Electrical Service, which has companies of electro-mechanics as its means of execution, is charged with providing for the general needs of electrical energy.

It establishes for this purpose installations of general interest of lights and electrical motive power as well as the organizations intended to supply them (makeshift power stations, distribution of the current coming from the large territorial power stations).

The electrical service of an army may be called upon by the high command to cooperate in the execution of large-scale installations involving several armies.

Forest Service.

493. The forest sapper companies, made up of mobilized forest chasseurs, are responsible for logging operations with the help of laborers and trains provided in principle by the staging units.

The conditions under which the forest resources of the Combat Zone are left at the disposal of the latter are fixed by specific instructions.

II. Engineering Park.

494. The engineer park includes:

— mechanized and pneumatic devices;

— matériel to repair road breaches;

— a reserve of matériels and tools.

III. Construction Directorate.

495. The works can be of different kinds depending on the general missions entrusted to engineers (No. 45):

— organization of the terrain (possibly organization of second and third positions);

— destruction;

— communication routes;

— installations of troops and services.

The directorate of these works is entrusted to the army to a senior engineering officer director of the works and their execution is ensured by:

— engineer units belonging to the army or belonging to the General Reserves placed at the disposal of the army;

— auxiliary laborers provided by the command to the body involved in the work.

ARTICLE 4.

SIGNAL COMMUNICATIONS SERVICE.

496. The matériel necessary for the Large Units or the *corps de troupe* is provided to them by the *signal communications parks* of the army placed under the orders of the commanders of the signal communications of the armies.

ARTICLE 5.

INTENDANCY SERVICE.

497. The Intendant-General, the senior chief of the Intendancy Service[1] of an Army, is a representative and delegate of the Army's commanding general and is responsible for the service in the Forward Zone and in the Communications Zone assigned to the Army, exercises the higher administration of the army intendancy units.

He submits to the command's approval all measures concerning the use and location of the intendancy units, supplies, evacuations of matériel, and the creation of stores or industrial organizations useful to the Army.

He inspects the intendancy bodies assigned in any capacity to the army and carries out all necessary checks.

---

[1] Roughly equivalent to the Quartermaster Corps in the U.S. Army or the Royal Army Service Corps in the British Army.—Trans.

He may be delegated by the Inspector-General of the Intendancy Service to inspect, as army supply units, the operation of the regulating station and the establishments of the interior and to verify the importance of the supplies that they contain.

He receives delegation of the credits necessary for all the services of the army and sub-delegates them as and when required to the directors of the services concerned, according to the instructions of the Commanding General of the Army.

The Directors of the Intendancy of the Army Corps and the Intendants of the Divisions have, in Army Corps and Divisions, roles analogous to that of the Intendant of the Army vis-à-vis the command, personnel, and intendancy units of their large unit.

498. The general operating rules of the Intendancy Service are as follows:

The supply service includes all the operations whose purpose is to gather and make available to the *corps de troupe* the supplies necessary for their subsistence and maintenance.

Supplies are provided either by shipments from the rear or by exploiting local resources.

The shipments from the rear come from the supply reserves maintained at a railroad station where the Minister has built up the necessary stocks that he puts, entirely or in part, at the disposal of the commander-in-chief.

Exploitation of the local resources of the Combat Zone is ensured by the Intendancy Service, according to the instructions of the command.

To meet the troops' needs at all times, varying-sized provisions reserves are set up at each echelon of command, which the supply service must methodically replenish as they are consumed.

Supply operations within armies essentially consist of the daily replenishing of the regimental trains of the *corps de troupe*.

For each Large Unit, these operations take place, except for meat, either at a supply railhead or at a supply depot.

Meat is transported by special motor vehicles to delivery centers where the *corps de troupe* has it removed by using their meat vehicles.

The supply of clothing is carried out under the same conditions as the supply of food. Clothing requested by the armies is transported from the interior clothing depots to the communication route regulators who ensure their delivery to the Large Units.

ARTICLE 6.

MEDICAL SERVICE.

499. The Medical Service Director of an Army is responsible for:

— to study and adopt, in accordance with the orders of the Army commander, all the measures concerning the medical service, in particular, the use and the location of the medical units of the Army and the administration of evacuations;

— supervising the maintenance and replenishment of matériel and regulating its distribution;

— prescribing and ensuring the execution of all hygiene and prophylactic measures.

The medical services directors of the Army Corps and Divisions, respectively, have, vis-à-vis the elements forming part of the Army Corps and the Division, or stationed in their zones, the same role and the same attributions as the Army's medical service director.

500. The functioning of the Medical Service in an Army is based on the following principles:

— concentration in the Combat Zone of the maximum means, in order to hasten as much as possible, the relief of the wounded and sending them to the rear to the specialized treatment centers;

— methodical triage of the sick and wounded, carried out at each echelon;

— rapid evacuation system enabling the injured to reach the medical facility in which they are to be treated in the shortest possible time;

— simple hospitalization system, organized in depth, based on the specialization of the formations and making maximum use of the resources of the Combat Zone.

The organization of evacuations and hospitalization is, at all echelons, the subject of a plan drawn up by the chief of the health service, according to the indications of the command and according to the existing resources in hospital resources, as well as traffic conditions on railroads and roads, waterways and, possibly, airways.

ARTICLE 7.

AERONAUTICAL SERVICE.

501. The Aeronautical Service (*service de l'air*) is responsible for ensuring for all Air Force (*Armée de l'Air*) formations:

— maintenance, troubleshooting, repair and disposal of matériel;

— the supply of matériels, liquid fuels and ammunition;

— the preparation and the equipment of the landing fields.

502. The supply of liquid fuels and ammunition is carried out as follows:

In agreement with the artillery service, the aeronautical service sets the points of delivery of the supplies in liquid fuels and ammunition.

It ensures by its own means (liquid fuel sections, aviation ammunition sections) the transport of these supplies from their delivery point to the landing field.

503. The Aeronautical Service of the air forces provide:

— to the air forces and D.C.A. command of the Armies (General Headquarters):

A Chief of the Aeronautical Service to the Armies;

— to the air forces and D.C.A. command of each Army:

A Chief of the Aeronautical Service of the Army, who has aeronautical service units of variable numbers (aerostation park, air park, aviation, specialized sections, airfield preparation companies, hangar fitters, etc.);

— the Air Forces Commanders of Army Corps (or Cavalry Divisions, fortified region, etc.), who perform the functions of chief of the aeronautical service for the formations under their command and possibly have units of the aeronautical service (supply sections, ballooning, etc.).

ARTICLE 8.

VETERINARY SERVICE.

504. The commander of each Large Unit is responsible for the upkeep and care of his animals; he decides on all the appropriate measures to maintain them in good condition. He receives, for this purpose, and if necessary, provokes all elicits proposals from the chief veterinarian of the service of the Large Unit.

The Chief of the Veterinary Service of each Large Unit is responsible vis-à-vis the command and his immediate technical superior, for the general operation of the veterinary service of the unit, the health surveillance of the animals, the treatment of the sick and wounded, evacuations, supplies of medicines and matériel. He has technical authority over the veterinarians of the formations making up the unit and gives them directives and controls their actions.

All animals likely to be cured in the short term are treated in the *corps de troupe*.

Contagious, overworked, seriously ill, and wounded animals are evacuated to Army veterinary hospitals by using mobile evacuation sections, divisions, and mobile evacuation groups.

Army veterinary hospitals operate under the supervision of the Veterinary Service Director of the Army.

ARTICLE 9.

REMOUNT SERVICE.

505. The Remount Service is organized by army under the direction of a chief of service (senior cavalry officer) who is part of the army headquarters and commands the mobile remount depot responsible for supplying the mobile remount groups.

The staffs, services, and *corps de troupe* are supplied with horses, mules, and pack animals by the army remount units, in principle by the mobile remount groups; however, the cavalry may also receive horses directly from their inland depots.

The chief of the remount service does not intervene in the distribution of animals from mobile remount groups among army units in deficit; he receives all useful instructions from the command (under the stamp of the 1st bureau of the general staff) in this respect.

ARTICLE 10.

POSTAL SERVICE.

506. The Armed Forces Postal Service is responsible for receiving and sending correspondence and postal parcels to the troops. Conversely, it carries correspondence and postal parcels from the Combat Zone to the interior. It also performs all other authorized postal operations.

All procedures must be implemented to ensure that postal correspondence to or from the armed forces reaches those concerned as soon as possible; the morale of the troops and that of the whole nation are greatly affected by it.

Only the Commander-in-Chief of the Armed Forces has the authority to delay the routing of correspondence from the armies, if he deems it necessary to ensure the secrecy of planned or ongoing operations.

The operation of the postal service is based on the allocation, to each division or grouping of similar importance, of a postal sector represented by a number and on the existence, within it, of a central military bureau receiving all the correspondence intended for the armed forces, classifying it by postal sector and sending it to its destination.

ARTICLE 11.

TREASURY SERVICE.

507. The Armed Forces Treasury Service is responsible for operating all receipts from the public treasury or made on behalf of the State, for paying all regularly ordered expenses for *corps de troupe* or services, and for carrying out all cash and treasury operations giving rise to the handling of funds and accounting.

Requests for the funds necessary for each pay office are established based on forecast statements drawn up by the intendancy officials for each ten-day period.

These funds are sent under escort by the Ministry of Finance to the paymaster of the regulating station, who forwards them to the paying offices concerned, either directly or through the army paymaster.

ARTICLE 12.

SERVICES OF FORTIFIED REGIONS.

508. A fortified region is provided with services comparable to those of an army corps.

The fortified region brigade has services comparable to those of an Infantry Division.

Certain autonomous fortified sectors are endowed with services that, apart from engineers, only include administrative bodies.

Each structure is equipped with all the services necessary for its own life. These are more or less developed depending on their importance and include in particular:

— repair shops and spare parts for artillery and engineering matériel;

— ammunition stores;

— engineer matériel depots;

— storage of food and health service matériel;

— an electro-mechanical service.

509. The operation, at each echelon, of the main services of the fortified organizations can be ensured without delay with the aid of special bodies constituted during peacetime or set up under the military services of the territory. From the moment the army services are in operation, the services of the fortified regions, brigades, and possibly sectors, continue to carry out their own missions, within the framework of the overall directives given by the Army commander responsible for the fortified region.

510. In the event of reinforcement by Large Units, the services of the fortified organizations that only correspond to specific needs (ammunition supply, electro-mechanical service) are juxtaposed with the services of these Large Units, with a view to avoiding at all costs the severing of the organic links existing in the regions or mixed brigades of the fortified region.

When a fortified region (or brigade) is divided between two armies, the commander of the region (or brigade) retains responsibility for regulating all the operations of the special services. He detaches representatives of his own services to the directors of the services of the two armies, if necessary.

Subject to these reservations, the general functioning of the services at each echelon is ensured by the authority exercising the tactical command of the sectors, whether it concerns the units assigned to the occupation of the works or those operating in the intervals. In particular, at the army level, it is up to the Commanding General of the Army to coordinate the various services as best as possible so as to take into account the above arrangements.

CHAPTER III.

TRANSPORTATION SERVICES.

ARTICLE 1.

GENERAL.

511. The army commander has, to ensure transport:

— organic formations of the army train, possibly reinforced by General Reserve formations;

— a credit (tonnage or manpower) allocated to the army by the Commander-in-Chief and taken from the General Reserve motorized formations, operated by the road regulatory commissions;

— possibly narrow-gauge railroad.

The army commander sets, in good time and generally every day, the transports to be carried out, their order of priority, and the deadlines within which they must be carried out.

He distributes these transports among the various means at his disposal, seeking the best use of each of these means to obtain the maximum overall output. In particular, he ensures that the transport units are only made available to a large subordinate unit or to a service on an exceptional basis, for a given mission, or for a specific period.

He establishes the army's *transportation plan.*

512. The army commander determines the road network necessary for the execution of movements and transport; he takes into account, in this designation, the routes on which the Commander-in-Chief is responsible for operating and supervising certain routes located in the zones of subordinate Large Units.

The army commander sets the rules for organizing traffic on the army road network thus defined, and the direction in which it must be improved. He establishes the army's *traffic plan.*

Traffic monitoring on the army road network is normally carried out by *road regulatory commissions*, bodies of the commander-in-chief, within the limits of their availability and following the instructions of the army commander. If necessary, these road regulatory commissions are reinforced by means placed at their disposal by the army commander.

Road traffic detachments organically belonging to motorized Large Units must, in principle, be left at the disposal of these large units for their own needs.

ARTICLE 2.

NARROW-GAUGE RAILROAD SERVICE.

513. The Director of the Narrow-gauge Railroad Service of an army is responsible for:

— maintenance and operation of existing narrow-gauge railroad networks in the Combat Zone;

— the study and construction of new networks to be created;

— maintenance of operating equipment.

Of capital importance in a period of stabilization when it constitutes a complete network, the narrow-gauge railroad service can still play a considerable role in mobile warfare. In this case, the efforts are concentrated on a limited number of branches. These branches are, as far as possible, chosen so as to intersect the normal railroad tracks at stations likely to be subsequently used as supply railhead stations; these stations are taken successively as the base of departure for transport by narrow-gauge railroads, as soon as the supply trains can arrive there.

ARTICLE 3.

TRAIN.

514. The Train Commander of an army provides transportation prescribed to him by the Commanding General of the Army.

He distributes this transport between the formations of the motorized train and the horse-drawn train placed under his orders, the horse-drawn train being, in principle, used over short distances or on roads unsuitable for motorized traffic.

The Train Commander of an Army Corps or Division ensures the execution of the transports that are prescribed to him by the Commanding General of the Army Corps or Division.

CHAPTER IV.

LAW AND ORDER SERVICES.

ARTICLE 1.

MILITARY JUSTICE SERVICE.

515. The Military Justice Service is assured by the military courts and by the military courts of cassation of the armies.

Cases submitted to military justice in the armed forces must be judged as quickly as possible, provided that the legal forms and deadlines are strictly respected.

It is indeed of the utmost importance that the punishment of crimes or misdemeanors, particularly acts of indiscipline, follow the misbehavior closely and that the accused soldiers remain away from the ranks for the shortest time possible.

ARTICLE 2.

MILITARY POLICE SERVICE.

516. At each echelon, the Military Police (*gendarmerie*) Commander (provost of the Large Unit) is placed under the orders of the chief of staff.

For the purposes of public order, policing, and traffic control, military police commanders have at their disposal the territorial gendarmerie units stationed in the Large Unit zone to which they belong.

CHAPTER V.

COMMUNICATIONS ZONE SERVICE.

517. The Director of Communications of an Army or of a Communications Zone of an Army Group or General Headquarters, commands and administers the territory placed under his authority. In national territory, he receives from the Army or Army Group Commanding General or from the Commander-in-Chief the directives relating to the relations to be maintained with the prefects and the sub-prefects for all questions concerning the army, in particular the possible needs of the civilian populations and their state of mind. In enemy territory, the General-Director of the Communications Zone is in charge of the provisional direction of the civil administration. An officer of the Communications Zone Service is placed near each local administrative authority which is, as far as possible, confirmed in its use. Employees and officials who appear suspicious are fired.

In any case, a general director of communications ensures the maintenance of order and the security of the lines of communication. He exploits the local resources according to the command's directives.

The forward limit of an Army Communications Zone is fixed by the Army's commanding general. Communication zones for Army Groups and headquarters are delimited by the Commander-in-Chief.

To ensure the exercise of his command and to carry out the missions assigned to him, a general director of communications has:

— a staff;

— service directors (engineering, aeronautics, intendancy, health, military police, military justice);

— a certain number of army communications commands where the main services are represented;

— line of communication troops and, possibly, units made available to him by the command.

TITLE XIII.

METHOD OF INSTRUCTION OF LARGE UNITS.

---

518. Large Units conduct their training with a view to:

— train the various services to lend each other the reciprocal support essential to success and to act in close liaison with the cooperation air forces and possibly with the navy;

— to operate, on the one hand, the staffs in conditions approaching those of war, on the other hand, the services in harmony with the operations;

— to obtain, at the different levels of the hierarchy, the psychological cohesion that comes from working together.

Troops receive their technical training in units below the division. The instruction of the Large Units aims especially at the instruction of the officers and noncommissioned officers (field-grade officers and signal communications specialists).

It includes exercises of various kinds, either on the map or on the ground.

As a general rule, the leader of the maneuvering unit exercises command of his unit; the leader of the higher unit leads the exercise.

The theme is established with a view to giving a specific lesson, either by studying a tactical situation with or without the participation of services, or by studying the sole functioning of one or more services. It must take into account the terrain and the psychological and matériel conditions in which the troops and the enemy find themselves very exactly: one will, therefore, choose the ground that best lends itself to the teaching to be given.

According to the matériel conditions in which the exercises take place, it is necessary to distinguish:

— officer exercises;

— troop exercises.

Officer exercises.

519. Officer exercises are the primary training process for Army Corps staff and services and higher units.

They are executed either on the map or on the ground, or by combining the two processes. This combination, which is particularly fruitful, allows officers to train in the following two operations which are common practice in the field:

— carry out a reconnaissance of the terrain after having methodically prepared it on the map;

— roughing out a tactical or technical decision using the map, then adjusting it definitively to the properties of the terrain.

The organization of signal communications is, in all cases, the subject of an in-depth study.

520. Officer exercises in the field advantageously take one of the following forms:

— officer exercises with troops reduced to the personnel and matériel essential to represent the various units in the field down to the elementary units (squads and pieces). In this case, only unit commanders take part in the exercise, with the automatic arms, weapons, or pieces of the corresponding unit and the personnel strictly necessary to serve them;

— exercise in the operation of command posts and signal communications enabling the various staffs and services to practice working under battlefield conditions.

The first of these processes allows the study of defensive situations more specifically, and the second of offensive situations.

Officer field exercises also include:

— reconnaissance in the frontier regions;

— staff rides.

521. Officer exercises are organized at each echelon within the framework of the higher unit, according to a frequency fixed by ministerial instructions.

For any officer exercise, the staff of the Large Unit concerned must be constituted as completely as possible. The services are either simply represented or composed in conditions approaching those of a war footing, depending on the purpose of the exercise.

Staffs must be trained to solve road traffic problems within the limited time available to them in the field.

The theme of each exercise must place the performers in front of a simple, soberly exposed situation, and should be sent to them sufficiently in advance so that they are in the exercise atmosphere from the start.

522. Exercises can be single or double action.

In single-action exercises, the exercise director must not let events unfold based on an undisturbed initial situation, but create incidents to force the various levels of command to make decisions.

In double action exercises, it is advisable, whenever possible, to maneuver one of the parties using the means and doctrine of the potential adversary.

Troop exercises.

523. Single- or double-action troop exercises are carried out:

— in the vicinity of garrisons (combined arms garrison or inter-garrison exercises);

— outside the camps;

— in the camps.

These three kinds of exercises are intended to study, under conditions that are increasingly close to reality, the *cooperation* of the various arms on the battlefield. They make it possible to develop and control the instruction of the command, staffs, and troops.

Exercises organized in the vicinity of garrisons pose more particularly *problems of a static character* (installation of a defensive position, preparation of a counterattack, deployment of an attack disposition on the base of departure). The cooperation of artillery, aviation, and armored vehicles, and the frequent participation of signal communications troops are sought first and foremost.

Maneuvers outside the camps allow the deployment of many Large Units and thus promote the instruction of the command, staffs, and services. They advantageously relate to situations characterized by *movement* (combat preliminaries and exploitation of success). They familiarize the troop with *life in the field* (walks, transport, encampment, security on the march and in the station, and supply).

Exercises carried out in the camps make it possible to present the synthesis of the fragmentary maneuvers carried out in garrison and outside the camps and to deal with the problems characterized by the *combination of fire and movement* (attacks).

They lend themselves to the overall study of defensive and, above all, *offensive* situations.

The full development of combat, including the use of live artillery and infantry fire, can be studied there. In any case, an effort is made to give the troops striking demonstrations of the effects of the fires of the various weapons.

524. Whatever terrain is used for the maneuver, all exercises with troops must be carefully prepared, conducted without haste, and only include the study of a very limited number of situations.

Moreover, they will only be profitable if a strongly constituted arbitration does not cease, while letting the executors exercise their initiative within the framework of the mission and the orders received, reminding them of the realities of the battlefield with all its constraints.

In this regard, it will be necessary to bring the maneuver to life in the eyes of officers and troops by embodying at all times the actions of the opposing artillery.

Approved,
Paris, August 12, 1936,
The Minister of National Defense and War,
Signed : Édouard DALADIER.

www.ingramcontent.com/pod-product-compliance
Lightning Source LLC
Chambersburg PA
CBHW021141090426
42740CB00008B/875

9798989320721